D0914071

Artificial Intelligence in Wireless Communications

For a listing of related Artech House titles,
turn to the back of this book.

Artificial Intelligence in Wireless Communications

Thomas W. Rondeau
Charles W. Bostian

Library of Congress Cataloging-in-Publication Data
A catalog record for this book is available from the U.S. Library of Congress.

British Library Cataloguing in Publication Data
A catalogue record for this book is available from the British Library.

Cover design by Igor Valdman

ISBN 13: 978-1-60783-234-8

© 2009 ARTECH HOUSE
685 Canton Street
Norwood, MA 02062

All rights reserved. Printed and bound in the United States of America. No part of this book may be reproduced or utilized in any form or by any means, electronic or mechanical, including photocopying, recording, or by any information storage and retrieval system, without permission in writing from the publisher.

All terms mentioned in this book that are known to be trademarks or service marks have been appropriately capitalized. Artech House cannot attest to the accuracy of this information. Use of a term in this book should not be regarded as affecting the validity of any trademark or service mark.

10 9 8 7 6 5 4 3 2 1

To my parents, Jo and Mike Rondeau – TWR

To my wife, Frieda Bostian – CWB

And to all of our friends, students, and colleagues for their insights, inspiration, and encouragement.

Contents

Acknowledgments

The work described here was supported in part by the National Science Foundation (Grants 9984363, DGE-9987586, and CNS-0519959) and the National Institute of Justice, Office of Justice Programs, U.S. Department of Justice (Grant 2005-IJ-CX-K017). Any findings, conclusions, or recommendations in this material are those of the authors and do not necessarily reflect the views of the National Science Foundation or the Department of Justice.

1

Introduction to Cognitive Radio

The current methods of communications are becoming less relevant under today's growing demand for and reliance on constant connectivity. Of decreasing relevance are the models of a single radio to perform a single task. The expansion of wireless access points among coffee shops, airports, malls, and other public arenas is opening up opportunities for new services and applications. In the data market, new technologies like the IEEE 802.16 WiMAX standard are being deployed, and mobile phone companies are offering services to customers for wireless connectivity over their networks without reliance on a WiFi access point. With current mobile phone access increasing worldwide and the Third Generation Partnership Project (3GPP) Long Term Evolution (LTE) project rapidly developing into a full standard, mobile phone companies are seriously competing in the wireless data market.

All of these applications and technologies offer trade-offs in quality of service and cost of service where quality of service is the effectiveness of the communications given requirements of time, data rate, form factor, and location. A person sending a text message on a train is not expecting an immediate response, but a conference call set up over a WiMAX connection demands real-time service. Because of these trade-offs, the market is responding by loading devices with more and more radios to allow consumers the ability to select their service depending on need and availability.

It is in these choices that we see the need for more intelligent decision-making from our devices. And this concept does not stop with just the choice between standards; the complexity of the standards themselves is growing. While WiFi devices manage their very few choices based on simple connection quality metrics, WiMAX and LTE include many waveform choices with highly complex interactions and consequences to the device behavior.

As both the applications for wireless services and the complexity of the devices increase, opportunities arise for more efficient ways to use and manage the wireless resources. The first major work being pursued here is the idea of dynamic spectrum access (DSA), which is technology that senses open channels and allows devices to communicate in underused parts of the spectrum. The idea of using intelligent signal processing and decision-making builds on this concept by enabling radios to manage not just spectrum but also the other available wireless resources. These radios would dynamically select spectrum, waveform design, time diversity, and spatial diversity options. They could even make changes at higher layers, for example, by modifying the medium access protocols or changing their routing behavior based on the network topology.

Radios that are capable of these intelligent decisions are called *cognitive radios*, and their actions are based on observing their wireless connections and then using intelligent algorithms and computational learning to optimize their behavior. Such actions modify aspects of different layers of the protocol stack for better performance as the current situation demands. To perform its activities, a cognitive radio applies a variety of techniques and capabilities developed in communications and networking research in a situationally dependent form. Where a mobile phone standard has addressed issues like operating at speeds under high multipath conditions, the standards-based solutions are trade-offs to provide the highest reliability among all of the possible conditions. These solutions will be suboptimal for many given situations. A better approach is to select the waveform and communications capabilities that work best for the given environment and situation.

In this book, we address the use of cognitive radio technology to provide communications systems with a specified quality of service by adapting the physical (PHY) and, to a small extent, Medium Access Control (MAC) layers. The major contribution is the formalization of radio optimization as a multiobjective optimization problem where radio resources are traded off to affect a desired quality of service driven either by a user or a specific application.

1.1 Brief Concept of Cognitive Radio

In discussing cognitive radio, we frequently talk about turning *knobs* and reading *meters*.[1] These terms come from classical transceivers with adjustable

1. To the best of our knowledge, our colleague Christian Rieser originated the terms "knobs" and "meters" as applied to cognitive radio in 2001 or 2002. They were quickly adopted by the research community and like other natural terminology, they may have arisen independently more than once.

controls (knobs) that determine the radio's operating parameters and meters that display certain performance or operating parameters of the system. An example is a broadcast frequency modulation (FM) receiver with a tuning knob to select which station to receive as well as equalizer knobs to adjust the sound quality. Meters might consist of a received signal strength measure or simply a light showing if the station is being received as mono or stereo sound.

We show a simple model of a cognitive radio in Figure 1.1 with the interaction between the cognitive engine and the radio through its knobs and meters. In cognitive radio terms, the waveform is the wireless signal transmitted that represents the current settings of all of the radio's knobs. Meters represent the metrics used in the radio optimization. Knobs include the type of modulation and modulation parameters, frequency channel, symbol rate, and channel and source coding. Meters include bit error rate (BER), frame error rate (FER), signal power, battery life, and computational resources.

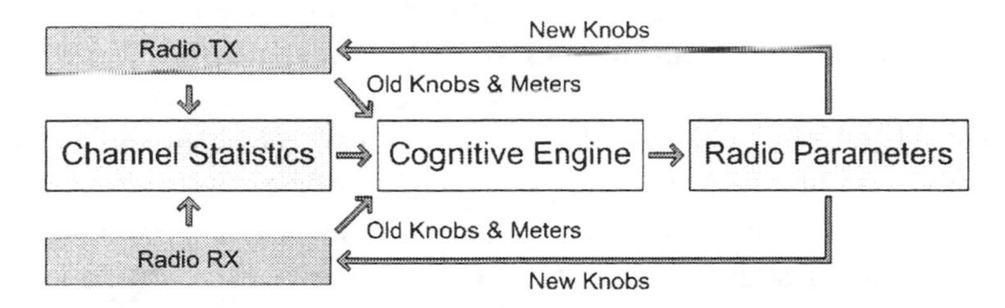

Figure 1.1 Simple model of a cognitive radio where the cognitive engine interacts with the radio through its knobs and meters.

Expressed in a single sentence, cognitive radio uses the meters to build an understanding of the environment so as to adjust the knobs to improve the communications.

1.2 Very Brief Cognitive Radio History

Early radios were designed with specific tasks in mind, like an FM radio or a television receiver for example. Even many contemporary devices operate in this way, such as public safety and WiFi networks. Mobile phones share many of these same features as they are normally dedicated to a single service, voice communications, but they are branching out more and more. Modern mobile phones generally support many different modes or waveforms for different networks and frequency bands as well as the ability to send text messages and, increasingly, data. Many are now equipped with Bluetooth and WiFi radios to extend their use and capabilities to different services. Other features include such techniques as adaptive power control or modulation adaptation in response to signal quality.

As both communications and computing technology advanced, it was inevitable that the two continue to integrate, defining the field known today as software defined radio (SDR). Communications devices are increasingly putting signal processing capabilities into software. As we discuss in more detail later, SDR provides many advantages and improvements in waveform design. With the flexibility SDRs offer, the next step was to utilize the computing power to adapt more of the waveform, making better use of the available communication system.

Joseph Mitola is credited with inventing the field of cognitive radios [1] with an interest in using the radio system as a personal assistant of sorts that intelligently reacts to the user's perceived needs. The concept of cognitive radios has since evolved towards a more communications-centric view of the radio. With a reconfigurable SDR system, a cognitive radio uses sensors to collect environmental information as well as an intelligent core to react to changes and challenges provided by the environment and user needs. A cognitive radio reacts and adapts to changes in the environment to provide continuous communications at a required quality of service (QoS).

In 2003 along with Christian Rieser and Bin Le at Virginia Tech, we successfully demonstrated a fully functional cognitive radio. We used a 5.8-GHz Proxim Tsunami radio system with a cochannel interferer. Our cognitive radio demonstrated streaming video. When the interferer interrupted the communications, the cognitive radio was able to develop a new waveform to overcome the interference. This work is described in detail in our SDR Forum paper [2].

For a comprehensive history and review of the goals of cognitive radio, the book *Cognitive Radio Technology* provides the first published collection of cognitive radio research [3]. Its chapters cover many different areas of cognitive radios, including history, policy and regulations, and implementation technology. Another comprehensive source of cognitive radio discussion is a set of papers published by Simon Haykin [4, 5]. We reference his work in the chapters to come. Other areas that are directly related to the implementation of cognitive radio and cognitive radio-like technology include the DARPA XG program, the IEEE 802.22 standard, and the IEEE P.1900 effort, now known as Standards Coordinating Committee (SCC) 41. The XG program [6] is a dynamic spectrum access (DSA) system that provides seamless communications while changing frequencies to keep from interfering with other networks while managing its own received interference.

The sensing, selection, and coordination of the use of radio spectrum, as well as a workable system, are all significant advances in the field. The goal of the IEEE 802.22 standard is to use cognitive radio technology to take advantage of unused spectrum for wireless regional area networks (WRAN) [7] while the IEEE SCC 41 works on more general cognitive radio and dynamic spectrum access standardization efforts [8].

The most significant and sustained work and research in cognitive radios comes from the field of DSA. Most of this work is chronicled in the proceedings of the IEEE DySPAN symposia [9, 10, 11]. The 2007 and 2008 symposia included a demonstrations section where companies and research groups were able to both demo their capabilities as well as experiment with other such devices. While DSA can be achieved without artificial intelligence, a lot of the research and solutions proposed used cognitive radio capabilities in their sensing and decision-making approaches.

The interest in DSA has been driven largely by the TV white spaces debates where the unused television channels could be repurposed for communications, which would require sophisticated sensing capabilities and frequency agility. The Federal Communications Commission (FCC) has recently published a positive report on testing such TV white-space sensing devices [12] and ruled on allowing such devices to operate [13], which is a major step forward for DSA and eventually cognitive radios. The FCC has not been silent on the subject in the past, either, issuing a report and order in 2005 on SDR and CR [14].

While we are still a few years away from commercial devices supporting the type of cognitive radios that this book discusses, the momentum is pushing in the right direction.

1.3 Definition

The definition of cognitive radio has been under debate since its introduction. In particular, much of the early work in cognitive radio dealt with the concept of DSA, and some authors equate cognitive radio with DSA. While this is one of the applications of cognitive radio, it is certainly not the only one. The other aspects of cognitive radio develop more of a service-oriented view of communications whereby the entire communications system adapts intelligently to offer better quality of service. The service model extends beyond the DSA model by looking at the system performance and not just the slice of spectrum allocated.

Instead of worrying about exact definitions as they are argued in the standardization bodies, the remainder of this book will deal with the goals of cognitive radio. These are to build a flexible, reconfigurable radio that is guided by intelligent processing to sense its surroundings, learn from experience and knowledge, and adapt the communications system to improve the use of radio resources and provide desired quality of service.

In this way, a cognitive radio follows traditional artificial intelligence systems [15]. These systems, as illustrated in Figure 1.2, act as agents that take input through sensors and respond to the input through actuators. The input to these systems are the radio's meters and the actuators are the radio's knobs. The intelligent agent completes the cognitive radio by providing the learning and intelligent algorithms that understand the meters and control the knobs.

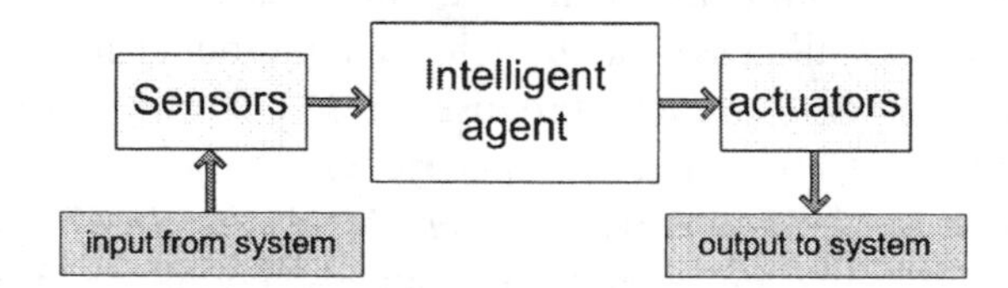

Figure 1.2 Typical artificial intelligent agent diagram that receives information through sensors and provides responses through actuators. When applied to a radio system, the agent provides the cognition to the cognitive radio.

1.4 Contributions

To enable the goals of cognitive radio, we discuss in this book an architecture of a cognitive engine to realize the necessary components of a cognitive radio. The cognitive engine, at a minimum, is designed to coordinate a set of sensors, an optimization routine, a learning and decision-making system, and the underlying reconfigurable radio system. We will show the design and implementation details of the cognitive engine components. Through the design of the cognitive engine, we discuss different applications of artificial intelligence to solve the problems faced by a cognitive radio and demonstrate how some of these methods improve communications.

The major theoretical work is in the detailing and discussion of the cognitive engine adaptation as a multiobjective optimization problem. For its application to cognitive radio, we describe the physical layer as a set of objective functions, methods of analyzing them, and ways that they can be traded off in the optimization system based on performance criteria. While this discussion provides the analysis only for the physical layer, we hope that it will show how to extend these techniques to other aspects of radio adaptation.

One particularly successful method to solve the multiobjective cognitive radio problems is the implementation of a highly flexible genetic algorithm. The operation of the genetic algorithm optimization system easily allows updates and additions to the optimization problem space as well as the dynamic creation of chromosomes to represent the waveform, and thus, it provides a solution independent of the search space and communications system.

As a way to augment the optimization process, we also introduce the use of case-based decision theory. This is a memory feedback system that learns and improves the cognitive radio behavior.

The cognitive radio design we will be discussing throughout this book covers all of these topics. We graphically summarize our approach in Figure 1.3 from the sensing to the monitoring feedback system.

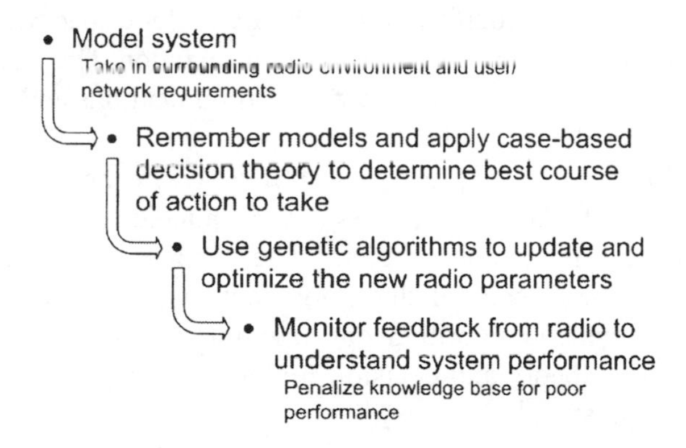

Model system
Take in surrounding radio environment and user/
network requirements
Remember models and apply case-based
decision theory to determine best course
of action to take
Use genetic algorithms to update and
optimize the new radio parameters
Monitor feedback from radio to
understand system performance
Penalize knowledge base for poor
performance

Figure 1.3 Overview of the major functions in our cognitive engine approach.

Although we provide both theory and a practical implementation of methods to build cognitive radios, we also understand that there is a lot of research left to be done to fully understand the cognitive radio behavior and optimization processes. Because of this, we have designed a mechanism that allows for easy introduction of new cognitive radio methods and components to test, experiment, and compare different solutions. The process also has the benefit of allowing easy distribution and sharing of cognitive components throughout a network, enabling knowledge sharing and distributed processing among cognitive radios.

We then provide experimental results of the cognitive engine on a real software radio platform. These experiments serve two purposes. First, they provide a scientific analysis of the theory we have developed. Second, we move the experiments between a simulated environment and real radios. Both of these systems are based on a similar underlying platform but with different capabilities. The cognitive radio methods are shown to easily transition between these two environments, supporting changing radio capabilities as well as performance objectives.

1.5 Contents

We begin this book by describing what a cognitive radio is and the pieces that, together, make a cognitive radio. The basic processing elements and their capabilities are implemented as modular components, and each component can be developed and tested independently before integration with the rest of the engine. This is discussed in detail in Chapter 2 and built upon throughout the rest of the chapters.

Contributions to the cognitive radio theory include specific implementations of artificial intelligence (AI) to radio optimization. Chapter 2 discusses the use of AI within the context of wireless communications and reviews some of the work that has developed here.

Chapter 3 introduces the principles of SDR. While there is no new theory added to the field of SDR, it is necessary to understand how they work in order to support the remaining chapters. The chapters on AI and SDR lead to the main theoretical focus of this book that includes radio optimization in Chapter 4, the genetic algorithm optimization method in Chapter 5, and case-based decision theory and decision-making in Chapter 6 that is used to improve the optimization.

Chapter 7 addresses the practical issue of controlling radio nodes in a network during reconfiguration of the physical layer waveform. Following these concepts of optimization and control with a cognitive engine, Chapter 8 provides a few working examples of the developed cognitive engine and experimental scenarios to understand the performance and behavior. Chapter 9 concludes the book by discussing a number of advanced topics and extensions to the theory and implementation provided here.

References

[1] J. Mitola, *Cognitive Radio: An Integrated Agent Architecture for Software Defined Radio*, PhD diss., Royal Institute of Technology, 2000.

[2] T. W. Rondeau, B. Le, C. J. Rieser, and C. W. Bostian, "Cognitive Radios with Genetic Algorithms: Intelligent Control of Software Defined Radios," in *Software Defined Radio Forum Technical Conference*, Phoenix, AZ, 2004, pp. C-3 – C-8.

[3] B. Fette, editor, *Cognitive Radio Technology*, New York: Elsevier, 2006.

[4] S. Haykin, "Cognitive Dynamic Systems," *IEEE Proc. Acoustics, Speech and Signal Processing*, Vol. 4, pp. IV-1369 – IV-1372, Apr. 2007.

[5] Simon Haykin, "Cognitive Radar: A Way of the Future," *IEEE Signal Processing Magazine*, Vol. 23, No. 1, pp. 30 – 40, Jan. 2006.

[6] M. McHenry, E. Livsics, T. Nguyen, and N. Majumdar, "XG Dynamic Spectrum Access Field Test Results," *IEEE Comm. Mag.*, Vol. 45, No. 6, pp. 51 – 57, June 2007.

[7] IEEE 802.22, 2007.

[8] IEEE 1900, 2007.

[9] IEEE. *IEEE Proc. DySPAN*, 2005.

[10] IEEE. *IEEE Proc. DySPAN*, 2007.

[11] IEEE. *IEEE Proc. DySPAN*, 2008.

[12] OET, "Evaluation of the Performance of Prototype TV-Band White Space Devices: Phase II," Technical Report, FCC, October 2008.

[13] FCC, "Second Report and Order and Memorandum Opinion and Order," Technical Report ET Docket No. 08-260, Federal Communications Commission, 2008.

[14] FCC, "Facilitating Opportunities for Flexible, Efficient, and Reliable Spectrum Use Employing Cognitive Radio Technologies: Report and Order," Technical Report ET Docket No. 05-57, Federal Communications Commission, 2005.

[15] M. Negnavitsky, *Artificial Intelligence: A Guide to Intelligent Systems*, Harlow, England: Addison-Wesley, 2002.

2

The Cognitive Engine: Artificial Intelligence for Wireless Communications

For a radio to become cognitive, we have to address many aspects in communications and computer science. We will also see how these concepts and their application to communications systems require a breadth of understanding of multiple disciplines. This makes cognitive radio a very exciting but also a very difficult research topic to study. Our intention in this chapter is to leap directly into the core framework of cognitive radio design before pulling back in subsequent chapters to some of the fundamental areas of research. In this way, we want to provide the big picture of cognitive radio research and design that paves the way for the importance of the fundamentals.

From our definition in Chapter 1, a cognitive radio is the application of intelligent processing and adaptation to a wireless communications system. We implement this as a processing engine to translate observations about the environment to action. We call this part of the radio a *cognitive engine*. In this chapter, we describe the function of a cognitive engine as well as the design of the cognitive engine we have developed. The description and development of the cognitive engine leads to a more general treatment of the use of learning techniques and artificial intelligence in wireless communications systems at the end of the chapter.

The cognitive engine (CE) is the intelligent core of the system described by Figure 1.2. Its design serves two simultaneous objectives:

1. Develop and apply cognitive radio algorithms;
2. Deploy cognitive radio functionality.

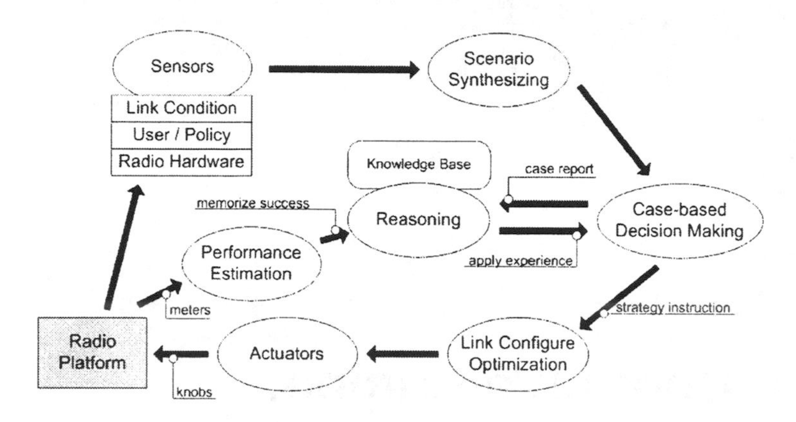

Figure 2.1 The cognition cycle with an outer loop for observation and optimization and inner loop for learning.

In particular, the actions of a cognitive radio (CR) follow the cognition cycle first proposed by Mitola [1]: observe, orient, plan, decide, learn, and act. A revision of the cognition cycle, published in a similar format in [2], is shown in Figure 2.1. This figure simplifies the actions of Mitola's loop and provides a system closer to a real implementation. In a way, it is a more detailed version of the AI agent of Figure 1.2. The radio platform provides both the input to the sensors and receives controls from the actuators. The sensors are labeled here as environment observations and cover a number of domains of information useful for a cognitive radio. The sensors pass the information on to the rest of the cognitive engine, which is a mixture of modeling, optimization, and learning algorithms. The outer loop contains a more straightforward synthesis of optimized waveforms while the inner loop provides long-term learning and reasoning abilities. The results of the cognitive engine are then used to control the radio by applying new knobs to create new waveforms. We use the theoretical aspects of this cognition loop to develop the cognitive engine architecture through this chapter.

2.1 Cognitive Radio Design

A cognitive radio is a flexible and intelligent radio capable of creating any waveform and using any protocol supported by the radio hardware and software. Waveforms consist of all of the parameters that define the way in which the radio transmits and receives information, including transmitter power, operating frequency, modulation, pulse shape, symbol rate, coding, and so forth. Protocols are the rules by which network nodes transfer information.

A cognitive radio develops waveforms and chooses protocols in real-time using artificial intelligence. These actions require three components:

1. Perception: Sensors that collect data on both external factors (channel conditions, other radios, regulations, user needs) and internal factors (waveform capabilities, available computational resources, remaining battery power).
2. Conception: An intelligent core that learns and understands how to combine knowledge from the sensing mechanism to aid the adaptation mechanism.
3. Execution: An optimization and adaptation mechanism that alters the radio's behavior.

Figure 2.2 presents a generic architecture for a cognitive radio. The cognitive engine is a separate entity within that radio that works alongside the normal communications path. The engine relies on information from the user, radio, and policy domains for instructions on how to best control the communication system. This structure works well as a generalized architecture as it makes no recommendations about how the cognitive engine (and therefore the rest of the cognitive radio) should behave while still mapping the interactions of the rest of the systems. The communications module itself appears as a simplified protocol stack, again showing the independence of the cognitive engine from the overall system.

Figure 2.2 shows three input domains that concern the cognitive radio. The user domain tells the cognitive engine the performance requirements of services and applications. Service and application requirements are related to the quality of service measures of a communications system. As each application requires different QoS concepts like speed and latency, this domain sets the performance goals of the radio.

The external environment and RF channel provide environmental context to the radio's transmission and reception behavior. Different propagation environments cause changes in the performance of waveforms that correspond to optimal receiver architectures. A heavy multipath environment requires a more complex receiver than simple line-of-sight propagation or log-normal fading. The external radio environment also plays a significant role in performance and adaptation. This environmental information helps provide optimization boundaries on the decision-making and waveform development.

Finally, the policy domain restricts the system to working within the boundaries and limitations set by the regulatory bodies as interpreted by the policy engine. The policy environment might determine a maximum amount of power a radio can use in a given spectrum or other spectrum rights with respect to other users, as was done in the 700-MHz band recently auctioned by

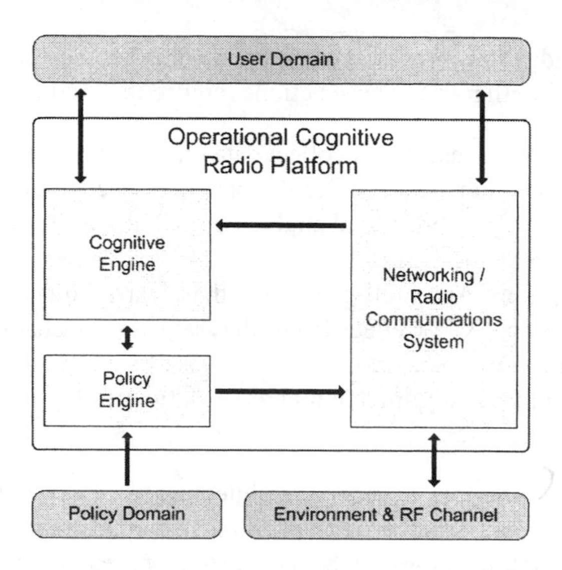

Figure 2.2 Generic cognitive radio architecture that receives input from three domains and controls a communications system.

the FCC [3]. Important regulatory action in the United States for our purposes includes the report and order on cognitive radio [4], recent action against open source software for software radios [5], and the regulations on Part 15 devices for use in unlicensed spectrum [6]. The rules from the FCC and other regulatory bodies impose constraints on the optimization space with respect to spectrum use and power.

2.2 Cognitive Engine Design

In order to better explain the operation of a cognitive engine, we now expand the discussion by introducing some specifics of our implementation.

To develop the cognitive capabilities of Figure 2.2, Figure 2.3 presents an architecture of the cognitive engine. It includes a central component called the cognitive controller that acts as the system kernel and scheduler to handle the input/output and timing of the other attached components. The other major components include:

Sensors: collect radio/environmental data.

Optimizer: given an objective and environment, create an optimized waveform.

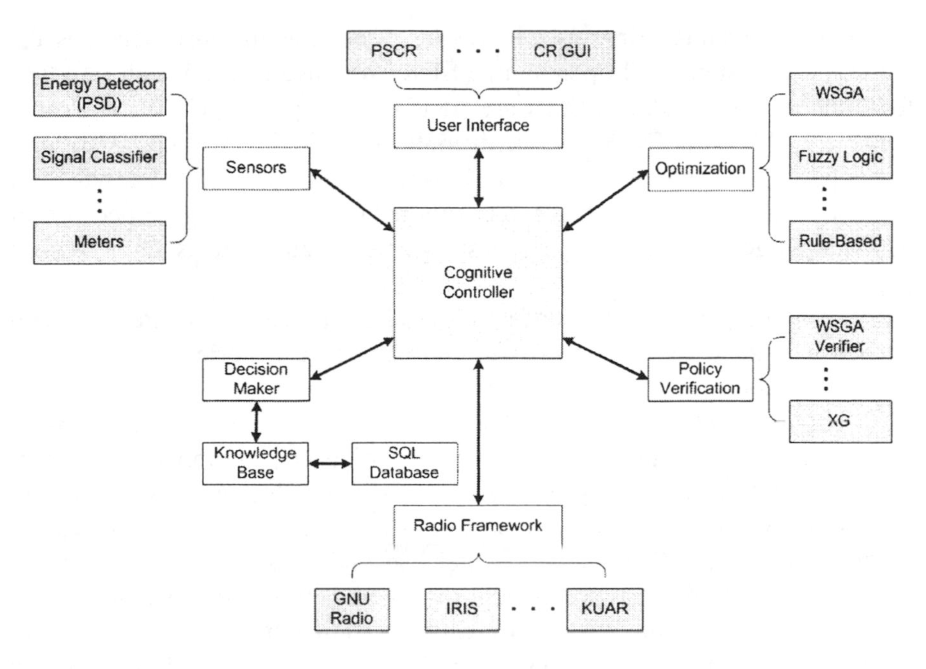

Figure 2.3 The cognitive engine developed at Virginia Tech's Center for Wireless Telecommunications (CWT) (patent no. 7289972).

Decision Maker: coordinate information and decide how to optimize and act.

Policy Engine: enforce regulatory restrictions.

Radio Framework: communicate with the radio platform to enable new waveforms and pull information from the sensors.

User Interface: provide control and monitor support to the cognitive engine.

In this figure, we point out some components that we have worked with and studied beyond what is covered in this book. One major point of the cognitive engine is to enable the integration of other algorithms and systems. For example, one user interface being developed at the Virginia Tech (VT) Center for Wireless Telecommunications (CWT) is a public safety cognitive radio (PSCR), which is an interface and control system for the cognitive engine designed for the public safety community [7].

The number of available SDR platforms is rapidly increasing. We have personal experience and working knowledge of a few for use with the cognitive engine. First, as we will be discussing in more detail, is the GNU Radio system, and open source SDR. We have also built an interface for Implementing Radio in Software (IRIS), developed by CTVR at Trinity College, Dublin [8]. The Kansas University Agile Radio (KUAR) is an SDR developed by the wireless team at Kansas University. While we have not developed an interface for this radio, we include it in the figure because of our knowledge of the radio system as a viable candidate for such a purpose [9].

Each component is launched as a separate process that interfaces and exchanges data with other processes through some generic interface (e.g., sockets or a message passing interface).

The architecture is designed around two important aspects. First, it allows development, testing, and launching of each component separately for low coupling between processes. This feature also enables distributed processing, where different components can reside on different processors or hosts with little change in behavior. Second, this architecture enables the testing of different types of algorithms and processes to realize each component. For example, many different sensors may be defined for different purposes and easily fit into the system, or different optimization functions may be developed and compared for performance. This architecture encourages both research and development.

2.3 Component Descriptions

The following sections provide more detail about the purpose and design of the system components.

Along with a standard interface to transfer information, the system also requires a standard for encoding the information. We chose eXtensible Markup Language (XML) as our method of conveying data and information. Using this standard satisfies a couple of competing goals. XML provides a method of encoding data that is open, flexible to support new and developing sensors, and both human and machine readable. XML also has a standardized format and methods of verifying the format, via a document type definition (DTD), to make integration with the cognitive controller easy. Finally, XML is an open, simple standard with many tools available to read and write it, including libraries for almost any programming language.

2.3.1 Sensors

Sensors collect data from the radio or other systems to describe and model the environment. Environmental data can include almost anything that will help the radio adjust its behavior, including radio propagation, interference models (temperature), position and location, time, and possible visual cues. The sensors collect the information by any means available or necessary: developed by a third party, prebuilt libraries, or specifically developed for use with the cognitive engine. In whatever manner the data is collected, the important aspect of a sensor is having a standard approach to how data is transferred to the cognitive controller. The application programming interface (API) is described as a simple state machine with a few important states:

- Initialization;
- Waiting for data request from cognitive controller;
- Collecting data and building model;
- Transferring model to cognitive controller.

Initialization builds the proper interfacing to the cognitive controller. The sensor then enters a wait state to listen to its interface for a request for data from the cognitive controller. When the sensor receives a request, it performs its data collection process, possibly by calling external libraries or applications, and then packages the data into an XML format to describe the sensor data. The XML data is transmitted to the cognitive controller, and the state machine returns to its wait state.

Another look at the structure of a sensor is shown in Figure 2.4. This figure shows that the cognitive engine sends information to the sensor through some generic interface. Sockets and Simple Object Access Protocol (SOAP) are two simple ways to communicate information between software programs either colocated on the same hardware or distributed and connected through a network layer. These are only two possible methods for providing the communications between the sensor and the engine. Functions and processing algorithms are retrieved through an external application or library. The use of an external interface here allows the sensor to access new or different functionality through simple and independent updates to a library or API.

Figure 2.5 lists the basics of the XML format for conveying sensor information to the cognitive controller.

The first line simply defines this as an XML v1.0 file. The next line says that this is a model from a sensor process. The particular model name is then the character data of the "model-name" tag. The remaining tags contain the model data. The important practice for proper representation of model data is shown in the fourth line where the data type, size, and unit are defined. The specification of the data type can be any type used by a particular database or language specific to the processing of the data. For instance, data used with a MySQL database can have the type "int," "float," or "char." The size information indicates the number of items included in this data tag; basically,

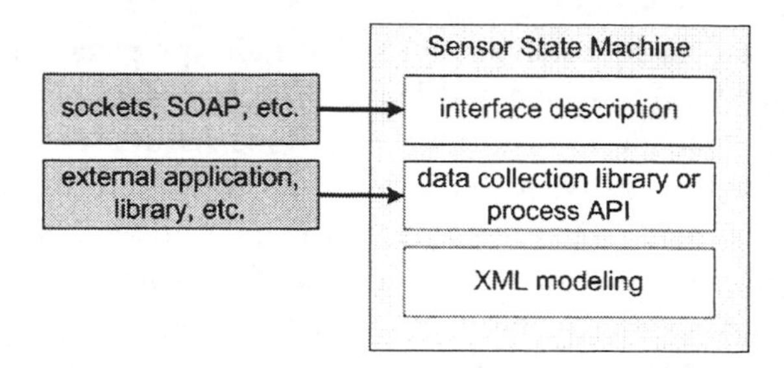

Figure 2.4 Sensor state machine architecture that receivers information from the cognitive engine.

```
<?xml version="1.0"?>
<sensor>
  <model-name>"model-name"<\model-name>
  <data-tag type="type" size="size" unit="unit">"value"<\data-tag>
  . . .
<\sensor>
```

Figure 2.5 Skeleton XML format for describing a sensor.

this is defined to help represent vector data. If this attribute is ignored, a value of "1" is assumed. The "size" attribute is an admittedly ugly method of enabling lists or vectors of data that are comma- or space-separated values. The final standard attribute describes the unit value the data represents, for example, dBm for power or Hz for frequency.

The script in Figure 2.6 below shows an example of a possible model received from a power spectral density (or energy detector) sensor. This example contains information on the system's noise floor at -85 dBm and one signal present in the environment that has a received amplitude of -50 dBm between the frequencies 449 to 451 MHz.

Perhaps the most important sensor is the sensor that collects meter information from the system, which we show in Figure 2.7. This sensor collects the system information such as noise power, signal power, bit error rate, battery life, or any other available meter. This information is very important to the optimization process, and lack of this information or wrong information could seriously impact the optimization process.

More information on these documents, format, and the initialization procedures are provided later as the cognitive engine is further developed.

```
<?xml version="1.0"?>
<sensor name="psd">
  <noise-floor type="double" size="1" unit="dBm">-85<\noise-floor>
  <signal>
    <amplitude type="float" size="1" unit="dBm">-50<\amplitude>
    <fmin type="float" size="1" unit="Hz">449e6<\fmin>
    <fmax type="float" size="1" unit="Hz">451e6<\fmax>
  <\signal>
<\sensor>
```

Figure 2.6 Example XML description of a PSD sensor.

```
<?xml version="1.0"?>
<sensor name="meters">
  <ber type="float" size="1">0<\ber>
  <per type="float" size="1">0<\per>
  <ebno type="float" size="1" units="dB">0<\ebno>
  <tx_signal_power type="float" size="1" units="dBm">0
                   <\tx_signal_power>
  <rx_signal_power type="float" size="1" units="dBm">0
                   <\rx_signal_power>
  <noise_power type="float" size="1" units="dBm">0<\noise_power>
<\sensor>
```

Figure 2.7 Example XML description of a meters sensor.

2.3.2 Optimizer

The optimization process takes environmental or user-oriented information from the sensors or user interface to select or design a waveform that will maximize the performance. Items that affect the optimization process include the user/application needs, the physical (propagation) environment, available resources (e.g., spectrum and computational resources), and the regulatory environment. Given a required QoS, the cognitive engine asks the optimizer to produce a waveform that comes as close to the QoS values as possible with respect to the provided environmental data. Depending on the implementation, the optimization may build a new waveform or select it from a list of predefined waveforms designed for specific problems.

The optimization process makes up a large part of this work and will be discussed in detail later, specifically in Chapter 5. Furthermore, there are many implementations of an optimization process with plenty of research in the field remaining. We touch upon some of these techniques later in this chapter, while the genetic algorithm approach presented in Chapter 5 provides a complete implementation as a starting point.

2.3.3 Decision Maker

The decision-making component of the cognitive engine helps understand the information provided by the sensors and helps make decisions about actions to take. The decision maker uses the sensor information to determine if reconfiguration is required due to poor performance or signs of decreasing performance. If optimization is required, the decision maker should also provide some context, such as an optimization goal (e.g., high throughput or low battery consumption) or a time limit for when a new waveform is required. The decision maker also uses past knowledge to provide the optimization process with information to help it in its work.

The current method of decision making in our work uses case-based decision theory (CBDT) [10]. CBDT keeps a database of observed cases, the actions taken to respond to those cases, and results of the action. When the sensor provides new data, the case that is the most similar and most useful is chosen from the case base as an action, or initial solution, to the optimization process. The decision maker then determines if optimization is needed to build a better waveform, use a waveform from the case base, or maintain the current waveform. If using a waveform from the case base, the cognitive radio could attempt to optimize the past solutions or bypass the optimization process altogether if that waveform performs well or if there is not enough time to find an alternative solution.

The decision maker and CBDT are discussed in detail in Chapter 6.

2.3.4 Policy Engine

The optimization process takes sensor data and creates a new waveform to meet some specified QoS. However, before the waveform returned by the optimization process can be sent to the radio, the cognitive controller must ensure it is legal with respect to the local regulatory restrictions. The policy engine does just that. The policy engine must test and authenticate a waveform. There are many ways to look at this process, but most of them involve databases of regulatory policies that restrict waveform transmission based on frequency and power with time as another possible dimension.

An important aspect of policy engines is that they must meet two competing goals within the cognitive radio world. First, the policy engine must be secure such that unauthorized waveforms cannot be transmitted. Second, it must be liberal enough to allow many different types of waveforms

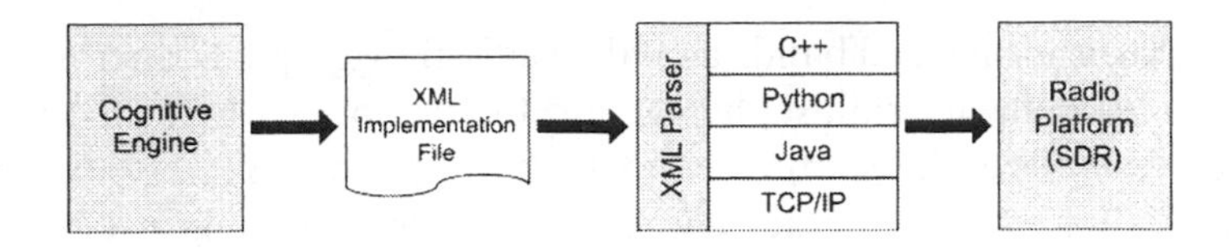

Figure 2.8 Translation process from generic XML format to radio-specific commands.

to run on the system as well as grow and change to match changing regulatory environments or waveform capabilities. Both of the above objectives require some form of authentication. The policy engine exists as an external component in the generic cognitive radio architecture of Figure 2.2 to help satisfy these requirements by allowing verified third-party systems to function here. We will not pursue the problem of policy and verification further. There is other work developing in this area, and we have pursued work such as [11] for our solution.

2.3.5 Radio Framework

The radio framework is the component that translates between the cognitive engine and the radio platform. This is effectively middleware between the generic representation used by the cognitive engine and the implementation-specific requirements of different radio systems.

When the cognitive engine wants to reconfigure the radio's waveform, it uses a generic, *communications theory* representation in XML that is most likely meaningless to the radio. The representation describes the physical layer behavior in terms of communications concepts like symbol rate, modulation type, and carrier frequency. The radio framework then translates these values to commands specific to the radio platform.

The mapping between the XML format to the radio-specific format is done by first parsing the XML file from the cognitive engine and formatting the commands used to configure the radio. The diagram in Figure 2.8 shows a generic interface for performing the translation with different modules plugged into the system for each different radio, as though they were device drivers. The XML Parser block is the translation block. It reads the XML format and converts to whatever format is required by the SDR. This could be a C++, Python, Java, or any other programming language API. The SDR control can also be accomplished by a more external interface such as through HTTP, message passing, or a proprietary interface.

The radio framework used in this work is the GNU Radio software radio. This radio is discussed in detail in Chapter 3. A simple Python XML parser reads the XML format and builds a GNU Radio flow graph. Scaperoth's paper [12] provides both the philosophy behind the use of XML for the interface language as well as the starting point to the waveform representation. The following XML shows a basic representation for describing a GNU Radio transceiver. The description is split into a hierarchy that describes the transmit and receive chains independently, and under each of these are branches to describe different parts of the communications stack, such as the physical and link layers. The values of the knobs in a particular layer are then defined in the leaves. It should be obvious how to add and define new knobs.

```
<?xml version="1.0" encoding="utf-8"?>
  <waveform type="digital">
    <Tx>
      <PHY>
        <rf>
          <tx_freq>408500000< \tx_freq>
          <tx_power>0.1< \tx_power>
        < \rf>
        <mod>
          <tx_mod type="PSK">
            <tx_mod_bits>1< \tx_mod_bits>
            <tx_mod_differential>1< \tx_mod_differential>
          < \tx_mod>
          <tx_rolloff>0.35< \tx_rolloff>
          <tx_bt>0.0< \tx_bt>
          <tx_gray_code>0< \tx_gray_code>
          <tx_symbol_rate>200000< \tx_symbol_rate>
        < \mod>
      < \PHY>
      <LINK>
        <frame>
          <tx_pkt_size>1450< \tx_pkt_size>
          <tx_access_code>0< \tx_access_code>
        < \frame>
      < \LINK>
    < \Tx>
    <Rx>
      <PHY>
        <rf>
          <rx_freq>408500000< \rx_freq>
          <rx_gain>35< \rx_gain>
        < \rf>
        <mod>
          <rx_mod type="PSK">
            <rx_mod_bits>1< \rx_mod_bits>
            <rx_mod_differential>1< \rx_mod_differential>
          < \rx_mod>
          <rx_rolloff>0.35< \rx_rolloff>
          <rx_bt>0.0< \rx_bt>
          <rx_gray_code>0< \rx_gray_code>
          <rx_symbol_rate>200000< \rx_symbol_rate>
        < \mod>
```

```
        < \PHY>
        <LINK>
            <frame>
                <rx_pkt_size>1450< \rx_pkt_size>
                <rx_access_code>0< \rx_access_code>
            < \frame>
        < \LINK>
    < \Rx>
< \waveform>
```

2.3.6 User Interface

The user interface has widely varying responsibilities depending on the cognitive radio use case. In one instance, it could be a control window where all actions and responses are controlled by a human operator, such as with a public safety radio. For more consumer-related applications where the cognitive radio should react autonomously and adapt based on the user and applications' requirements, the user interface may be a simple configuration window setting up certain parameters. In the most idealistic view of cognitive radios, there is no user interface and the cognitive engine simply acts on its own.

2.3.7 Cognitive Controller Configuration

An important aspect of the cognitive controller is its ability to use many different implementations of the components described above. To enable this capability, each component is defined around a basic state machine that interfaces between the controller and the component. The cognitive controller, then, is configured through an XML file that defines which components are currently attached, as shown below. The interfacing can be defined as any potential transport layer. For example, in the current design, simple TCP sockets are used and defined by the hostname of the system running the component and the port number the component is listening to. This design makes it simple to distribute processes among networked nodes just by changing the hostname. To do this, of course, the transport needs to be secure and stable.

```
<?xml version="1.0" encoding="utf-8"?>
<cognitive-controller>
        <knowledge-base>
                interface information
        < \knowledge-base>
        <sensor>
                <name>meters< \name>
                interface information
```

```
        < \sensor>
        <sensor>
            <name>psd< \name>
            interface information
        < \sensor>
        <optimizer>
            interface information
        < \optimizer>
        <radio>
            interface information
        < \radio>
        <user-interface>
            interface information
        < \user-interface>
        <policy-engine>
            interface information
        < \policy-engine>
< \cognitive-controller>
```

In this listing, each type of component is defined. Because a cognitive radio will likely have multiple input methods for gathering information, the cognitive controller can define and connect to multiple sensors. Here, the cognitive radio has a sensor to collect the PSD of the radio environment as well as a sensor that collects radio performance meters. Each is described by a specific name that the cognitive controller uses to identify the sensor when collecting information.

2.4 Artificial Intelligence in Wireless Communications

Successful cognitive radios are aware, can learn, and can take action for any situation that might arise. Applications range from voice communications under low power conditions to communications in high interference zones to more complex, critical, and hostile military networks of interoperating vehicles and soldiers with many different network needs. A radio must respond to any of these scenarios and adapt the many different parameters that define its waveform and protocols. These radios do not just require learning; instead, they need highly sophisticated learning and decision-making capabilities.

Machine learning has been well documented and received both criticisms [13] and praise [14]. Successful applications of AI are often limited to narrowly defined, well-bounded applications. While waveform adaptation is a bounded problem, the technical demands for intelligence in a radio exceed those normally associated with successful applications of classic artificial intelligence techniques, such as expert systems or neural networks. Waveform optimization requires stronger reasoning capabilities and the potential to create and test new design solutions.

A common theme we will continue to develop is the combined use of both learning and optimization processes. Feedback from a learning system can augment the optimization routines through comparisons between the radio's actions and the desired outcomes of the optimization. Furthermore, as touched upon in this chapter and developed in Chapter 6, a learning system can significantly aid decision-making in time-constrained situations. If the cognitive radio requires an immediate solution, the learning system can provide a known working solution developed in the past. Given time constraints or lack of valid solutions, the optimization process can develop new solutions or evolve old solutions for better operation.

Information and knowledge are both important concepts for a cognitive radio. Information is data about the environment collected through the available sensors. Information can include such items as position, interference, battery life, or performance analysis. The information collected from the sensors feeds both the learning and the optimization routines to help them make decisions. Knowledge is a concept developed from information. Knowledge is a useful representation of the information that says something about what the information means. The sensors might provide the cognitive radio with time and position information, but the radio needs to know what that information might mean about potential use patterns and known problems, such as areas of service outage or high interference during a daily commute.

More information is good, but only if the cognitive engine can transform the information into usable knowledge. Some sensors might provide a lot of information such as ambient temperature, but if the models used to make decisions do not use that information, the sensor adds no useful knowledge to the system. On the other hand, sophisticated sensors that provide information about interference power over a wide bandwidth can find immediate use by a cognitive radio seeking access to a particular amount of spectrum.

2.5 Artificial Intelligence Techniques

Below, we list several AI techniques/areas receiving considerable attention in the literature on cognitive radio. We present and review a few particularly relevant papers regarding each AI technique, specifically the papers that well represent the field or that provide comprehensive background themselves.

There are a couple of well-known areas of AI techniques that we purposefully leave out. One large technique in signal processing is the Bayesian network. This powerful learning technique based in Bayes' theorem uses past experience to enhance future decisions. The reason this technique is excluded from the following discussion is because there is little published work in the use of Bayesian networks in cognitive radio. Haykin, who has a history of work with these and other AI techniques in communications, cites the approach in his article on cognitive radar [15], but does not offer details for how to employ it. Brief discussions of Bayesian networks appear in a few publications, and they are used in other aspects of communications and signal processing, and so the technique will likely start making a serious impact soon.

Another popular AI technique is the expert system. Expert systems have been successful in some applications, particularly early in the development of AI, such as the DENDRAL project in organic chemistry [16]. Mitola addresses the concept of expert systems at length in his dissertation on cognitive radio [17] in which he mentions the ideas of "knowledge-engineering bottlenecks and software of limited flexibility." The bottlenecks occur due to the need for domain experts to define all knowledge, action, and behavior of the expert system. This same principle also limits any refinement of the expert system without further relying on such experts, thereby limiting how flexible the system is to new situations and in the face of new constraints and information. There may exist limiting cases for the use of expert systems, but it is not an independent approach to realizing cognitive radio.

The next few sections highlight different artificial intelligence techniques that use information to make knowledgeable decisions in the cognitive radio. As Arthur C. Clarke famously said, "any sufficiently advanced technology is indistinguishable from magic." Likewise, it might be true that any sufficiently advanced signal processing algorithm is indistinguishable from artificial intelligence.

2.5.1 Neural Networks

Neural networks are among the oldest form of AI in computer science, starting with the mathematical formulation by McCulloch and Pitts [18]. They have come and gone as a fad over the decades, but recent advances, both hardware and software, enable their use in more applications. Of particular importance to cognitive radios, neural networks provide a means for signal and modulation detection and classification.

Chan et al. [19] did some of the early published work on signal classification algorithms with decision theoretic and pattern matching. Both methods used time-based statistics, and neither proved too robust under low SNR conditions. Azzouz and Nandi then did some important work on the subject [20] and did some of the early work using neural networks as the signal processing technique of choice that showed greater promise in classification of signals under noisier conditions [21]. The use of neural networks in modulation classification has since become a well-accepted technique using both time-based statistics [22] and frequency analysis [23] as the inputs to the network.

Neural networks are really signal processing elements that perform simple operations on data. However, the collection of artificial neurons and clever learning algorithms allow networks to build and adapt to represent and process data in interesting ways. In signal classification, they take multiple noisy input items and provide accurate answers to the type of modulation represented.

2.5.2 Hidden Markov Models (HMM)

In some circles, hidden Markov models (HMM) [24] might be considered artificial intelligence, though we certainly would not categorize them as such. A HMM is a processing tool that uses past data to help predict future actions; an implementation of Bayes' law. We discuss them here because they are useful in communications and cognitive radios.

The best reference to learn about how HMMs work is Rabiner's tutorial [24]. Channel modeling has extensively used Markov models in research. Probably the most famous is the two-state Gilbert-Elliot model [25] that describes a channel as in either a good state or bad. When in one state, there is a probability of either staying in that state or moving to the other state. The channel properties determine the type of transition probabilities. Researchers have developed other, more extensive models, and [26] provides a good comprehensive overview of these.

The idea of developing such a model lends itself to cognitive radios. Rieser and Rondeau looked into using HMMs in channel models where they used a genetic algorithm as the training method instead of the Baum-Welch algorithm [27]. The HMM in this instance was used to provide a compact channel model based on information gathered in a live system to represent the current channel statistics. The idea was to use the HMMs as a sensor to understand the channel behavior in a cognitive engine, although the research was not taken much farther in this direction.

Mohammad's work used HMMs for a similar purpose, but was able to develop classification schemes in order to use the models for decision making in a cellular network [28]. The ability he developed to calculate a similarity distance between HMMs provides promise for future implementation in a cognitive radio system, especially in the context of the environmental modeling used in a case-based system as discussed in Chapter 6.

2.5.3 Fuzzy Logic

Fuzzy logic is a famous technique that started during the early development of artificial intelligence [29, 30]. Because it deals extensively with uncertainty in decision making and analysis, it has great potential for application to cognitive radio. However, only a little work has so far been published in the field, notably by Baldo and Zorzi [31]. Their implementation suggests some interesting applications, and the discussion points out larger uses than the specific application of adapting the TCP layer used in the paper. A problematic aspect of this work is the amount of domain-specific rules required. All implementations of AI require domain information, but fuzzy logic must establish a rule related to the specific situation in which it is used. Programming these rules recalls some of the limitations of expert systems, although fuzzy logic is still far more flexible and powerful. Fuzzy logic has potential in either specific problem solving areas or as a part of a cognitive radio.

2.5.4 Evolutionary Algorithms

Along with Christian Rieser, we pioneered the use of genetic algorithms [32, 33] early in cognitive radio research [34, 35, 36], which this work extends. The basic principles, as discussed throughout, are that the large search space involved in optimizing a radio is more complex than many search and optimization algorithms can handle. Among those algorithms that are suited to the task, evolutionary, specifically genetic, algorithms offer a significant amount of power and flexibility. Cognitive radios are likely to face dynamic environments and situations as well as radio upgrades due to advancing technology, so the flexible representation of the problem space allowed genetic algorithms are particularly applicable.

More recently, Newman et al. [37] have also contributed significantly to the use of genetic algorithms for cognitive radios. As we discuss in detail in Chapters 4 and 5, one of the main issues involved in successful genetic algorithm behavior is the selection of the fitness, or objective, function(s). Newman's work has developed a single, linear objective function to combine the objectives of BER minimization, power minimization, and throughput maximization.

2.5.5 Case-Based Reasoning

The final traditional AI technique to discuss here is case-based reasoning (CBR) [39]. CBR systems use past knowledge to learn and improve future actions. In these systems, a case base stores actions and receives inputs from a sensor. Those inputs help find the action in the case base that best fits the information received by the sensor. As mentioned previously, an optimization routine could, instead of designing a new waveform, select a waveform from a predefined list. CBR is a method used to make the associations. Although this may sound like an expert system, CBR systems generally provide learning and feedback to continuously and autonomously improve their performance. As information is received and actions taken, the results can help the system improve its response the next time.

Another contribution from Newman et al. [37] develops a similar idea in the experiments they run using previous knowledge to seed the next run of the genetic algorithm. The cognitive radio remembers solutions found for one particular problem to apply to the next problem to initialize the population with known successful chromosomes. The population seeding in [37] resembles the case-based decision theory work presented in Chapter 6. Their seeding concept uses a factor to calculate the expected change in the environment between runs of the genetic algorithm to provide context for how successful a new chromosome might be with respect to the new environment. We will show later how the case-based work extends this idea by keeping a set of previously observed cases and finding which case best matches the current environment as opposed to assuming certain changes in the environment.

2.6 Conclusions

In this chapter, we introduced the concept of the cognitive engine and began to show the implementation we developed to realize the structure in an extensible, flexible platform. The major components of the platform include sensors, optimizer, decision maker, policy engine, radio framework, and the user interface. The discussion of this chapter focused mostly on defining the roles and responsibilities of each component to provide the context from which to build a cognitive radio. From this initial discussion, we now begin developing the theory that enables the different parts of the engine. We will then revisit the engine's design in much greater detail.

To realize a cognitive radio, AI provides many viable techniques and tools. We have discussed many of these techniques with a brief literature review of each as related to their application in cognitive radio. In later chapters, we use and develop these ideas more deeply in the design of the cognitive engine framework presented here.

Our work deals largely in the optimization routine in Chapters 4 and 5 and on the decision maker and learning routine in Chapter 6. From this introduction of the cognitive radio and cognitive engine, the next chapter introduces the radio framework required for use by the AI approaches in the later chapters.

References

[1] J. Mitola and G. Q. Maguire, Jr., "Cognitive Radio: Making Software Radios More Personal," *IEEE Proc. Personal Communications*, Vol. 6, 1999, pp. 13 – 18.

[2] T. W. Rondeau, C. W. Bostian, D. Maldonado, A. Ferguson, S. Ball, B. Le, and S. Midkiff, "Cognitive Radios in Public Safety and Spectrum Management," *Telecommunications Policy and Research Conference*, Vol. 33, Sep. 2005.

[3] FCC, "Implementing a Nationwide, Broadband, Interoperable Public Safety Network in the 700 MHz Band," Federal Communications Commission, Tech. Rep. PS Docket No. 06-229, Dec. 2006.

[4] ——, "Facilitating Opportunities for Flexible, Efficient, and Reliable Spectrum Use Employing Cognitive Radio Technologies: Report and Order," Federal Communications Commission, Tech. Rep. ET Docket No. 05-57, 2005.

[5] ——, "Cognitive Radio Technologies and Software Defined Radios," Federal Communications Commission, Tech. Rep. ET Docket No. 03-108; FCC 07-66, 2007.

[6] ——, "Title 47 of the Code of Federal Regulations: Part 15-Radio Frequency Devices," Federal Communications Commission, Tech. Rep., 2001.

[7] B. Le, P. Garcia, Q. Chen, B. Li, F. Ge, M. ElNainay, T. W. Rondeau, and C. W. Bostian, "A Public Safety Cognitive Radio Node System," *Software Defined Radio Forum Technical Conference*, Denver, Colorado, 2007.

[8] P. Sutton, L. Doyle, K. E. Nolan, "A Reconfigurable Platform for Cognitive Networks," *IEEE Proc. Cognitive Radio Oriented Wireless Networks and Communications (CROWNCOM)*, Jun. 2006, pp. 1 – 5.

[9] J. D. Guffey, A. M. Wyglinski, and G. J. Minden, "Agile Radio Implementation of OFDM Physical Layer for Dynamic Spectrum Access Research," *IEEE Proc. GLOBECOM*, Washington, D.C., Nov. 2007, pp. 4051 – 4055.

[10] I. Gilboa and D. Schmeidler, *A Theory of Case-Based Decisions*, Cambridge: Cambridge University Press, 2001.

[11] P. Cowhig, "A Complete & Practical Approach to Ensure the Legality of a Signal Transmitted by a Cognitive Radio," Masters thesis, Virginia Tech, 2006.

[12] D. Scaperoth, B. Le, T. W. Rondeau, D. Maldonado, C. W. Bostian, and S. Harrison, "Cognitive Radio Platform Development for Interoperability," *MILCOM*, Washington, D.C., Oct. 2006, pp. 1 – 6.

[13] J. Hawkins, *On Intelligence*, New York: Times Books, 2004.

[14] M. Negnavitsky, *Artificial Intelligence: A Guide to Intelligent Systems*, Harlow, England: Addison-Wesley, 2002.

[15] S. Haykin, "Cognitive Radar: A Way of the Future," *IEEE Signal Processing Magazine*, Vol. 23, No. 1, pp. 30 – 40, Jan. 2006.

[16] B. G. Buchanan, G. L. Sutherland, and E. A. Feigenbaum, "Heuristic DENDRAL: A Program for Generating Explanatory Hypotheses in Organic Chemistry," *Machine Intelligence*, Vol. 4, pp. 209 – 254, 1969.

[17] J. Mitola, "Cognitive Radio: An Integrated Agent Architecture for Software Defined Radio," Ph.D. diss., Royal Institute of Technology, 2000.

[18] W. McCulloch and W. Pitts, "A Logical Calculus of Ideas Immanent in Nervous Activity," *Bulletin of Mathematical Biophysics*, Vol. 5, pp. 115–133, 1943.

[19] Y. Chan, L. Gadbois, and P. Yansounix, "Identification of the Modulation Type of a Signal," *IEEE Proc. Acoustics, Speech, and Signal Processing*, Vol. 10, Apr. 1985, pp. 838 – 841.

[20] E. E. Azzouz and A. K. Nandi, "Procedure for Automatic Recognition of Analogue and Digital Modulations," *IEE Proc. Communications*, Vol. 143, No. 5, pp. 259 – 266, Oct. 1996.

[21] A. K. Nandi and E. E. Azzouz, "Algorithms for Automatic Modulation Recognition of Communication Signals," *IEEE Trans. Communications*, Vol. 46, No. 4, pp. 431 – 436, Apr. 1998.

[22] B. Le, T. W. Rondeau, D. Maldonado, D. Scaperoth, and C. W. Bostian, "Signal Recognition for Cognitive Radios," *Software Defined Radio Forum Technical Conference*, 2006.

[23] A. Fehske, J. Gaeddert, and J. H. Reed, "A New Approach to Signal Classification Using Spectral Correlation and Neural Networks," *IEEE Proc. DySPAN*, 2005, pp. 144 – 150.

[24] L. Rabiner, "A Tutorial on Hidden Markov Models and Selected Applications in Speech Recognition," *Proc. IEEE*, Vol. 77, No. 2, pp. 257 – 286, Feb. 1989.

[25] E. N. Gilbert, "Capacity of a Burst-Noise Channel," *Bell Labs Technical Journal*, Vol. 39, pp. 1253 – 1266, Sep. 1960.

[26] L. N. Kanal and A. R. K. Sastry, "Models for Channels with Memory and Their Applications to Error Control," *Proc. of the IEEE*, Vol. 66, No. 7, pp. 724 – 744, Jul. 1978.

[27] T. W. Rondeau, C. J. Rieser, T. M. Gallagher, and C. W. Bostian, "Online Modeling of Wireless Channels with Hidden Markov Models and Channel Impulse Responses for Cognitive Radios," *IEEE Proc. IMS*, 2004, pp. 739 – 742.

[28] M. Mohammad, "Cellular Diagnostic Systems Using Hidden Markov Models," Ph.D. diss., Virginia Tech, 2006.

[29] L. A. Zadeh, "Fuzzy Sets," *Information and Control*, Vol. 8, pp. 338 – 353, 1965.

[30] M. Black, "Vagueness: An Exercise in Logical Analysis," *Philosophy of Science*, Vol. 4, No. 4, pp. 427 – 455, Oct. 1937.

[31] N. Baldo and M. Zorzi, "Fuzzy Logic for Cross-Layer Optimization in Cognitive Radio Networks," *IEEE CCNC*, Jan. 2007, pp. 1128 – 1133.

[32] D. E. Goldberg, *Genetic Algorithms in Search, Optimization, and Machine Learning.* Reading, MA: Addison-Wesley, 1989.

[33] J. Holland, *Adaptation in Natural and Artificial Systems*, Boston: MIT Press, 1975.

[34] C. J. Rieser, "Biologically Inspired Cognitive Radio Engine Model Utilizing Distributed Genetic Algorithms for Secure and Robust Wireless Communications and Networking," Ph.D. dissertation, Virginia Tech, 2004.

[35] C. Rieser, T. Rondeau, C. Bostian, and T. Gallagher, "Cognitive Radio Testbed: Further Details and Testing of a Distributed Genetic Algorithm Based Cognitive Engine for Programmable Radios," *IEEE Military Communications Conference*, Nov. 2004.

[36] T. W. Rondeau, B. Le, C. J. Rieser, and C. W. Bostian, "Cognitive Radios with Genetic Algorithms: Intelligent Control of Software Defined Radios," *Software Defined Radio Forum Technical Conference*, 2004, pp. C–3 – C–8.

[37] T. R. Newman, R. Rajbanshi, A. M. Wyglinski, J. B. Evans, and G. J. Minden, "Population Adaptation for Genetic Algorithm-Based Cognitive Radios," *IEEE Proc. Cognitive Radio Oriented Wireless Networks and Communications*, Aug. 2007.

[38] P. Mähönen, M. Petrova, J. Riihijarvi, and M. Wellens, "Cognitive Wireless Networks: Your Network Just Became a Teenager," *IEEE INFOCOM*, 2006.

[39] J. Kolodner, *Case-Based Reasoning*, San Mateo, CA: Morgan Kaufmann Pub., 1993.

3

Overview and Basics of Software Defined Radios

One of the most important enabling technologies to the application and success of cognitive radios is an adaptive, flexible, and powerful radio platform. Many cognitive radio techniques can work with more traditional radio designs. For instance, new mobile phones and laptop computer chips are offering access to multiple radio technologies and services. Such service selection can be accomplished by cleverly designing RFICs and software to take advantage of different standards like WiFi, WiMAX, LTE, CDMA, and GSM. A simple, but useful, cognitive radio implementation would use multistandard technology to automatically select the carrier and standard that best provides the QoS and pricing for the current application. We present a method to do this in Chapter 6.

More interesting applications and cognitive radio capabilities become evident when we break away from traditional, fixed standards-based radios and move to more flexible platforms. Such flexibility is coming about by moving the signal processing into the software domain. Device capabilities are easily adjusted and turned on and off through software functions, and software updates provide paths to introduce new and better services. With the added flexibility, radios can then determine and negotiate their own waveforms as needed to satisfy their service requirements. In this chapter, we introduce software radios. The rest of the book will take the basics we develop here as the context for building cognitive radio applications.

The first part of this chapter covers a basic introduction to SDR or software radio (SR). We will not get into any notational argument between SDR and SR and will use them interchangeably. We will not provide an in-depth analysis of the subject of software radio, but we will present the necessary information on SDR technology to explain why it is used as the implementation platform for cognitive radio. Within this context, it is important to point out both the benefits and potential problems with SDR. For a more complete coverage of SDR technology as well as a comprehensive list of references in the field, see Reed [1] and Tuttlebee [2].

In the second half of this chapter, we present a discussion of the *GNU Radio* software radio. This is an open-source SDR implementation that provides many benefits to the cognitive radio researcher. It is free, open and transparent, and increasingly powerful as new capabilities and updates are introduced to the core system. This information is intended to provide the reader with an understanding of how GNU Radio, as a specific software radio implementation, can be used as the platform for cognitive radio experimentation.

3.1 Background

As the name suggests, a software defined radio is a radio system where the majority of physical layer signal processing is done in software. The signal processing encompasses modulation, forward error correction, spreading, filtering, phase, frequency, timing synchronization, and so on.

Figure 3.1 shows the concept of an *ideal* SDR where the received signal comes in from an antenna, is converted to the digital domain via the analog to digital converter (ADC), and the rest of the signal processing is done in software. Likewise, the transmitter performs all the signal processing in software and sends the signal out of the antenna via the digital to analog converter (DAC). Unfortunately, in this type of system, the requirements of the ADC and DAC as dictated by dynamic range, sampling rates, and bandwidth specifications far exceed practical capabilities (see [3] for details of ADC technology). Likewise, running software instead of hardware implementations of communications systems introduces performance limitations.

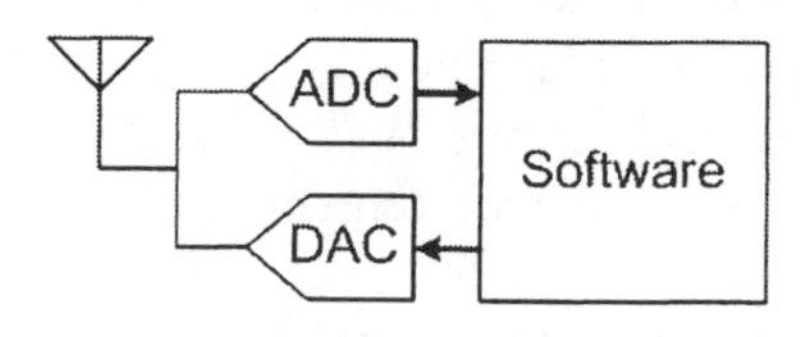

Figure 3.1 Ideal SDR with direct conversion between analog and digital world at the antenna port.

Given the limitations of realizing the ideal SDR, hardware can perform some of the signal processing while processors of different types, such as field programmable gate arrays (FPGA), digital signal processors (DSP), and general purpose processors (GPP) can handle other parts. Figure 3.2 shows a very high-level view of a radio transmitter that includes some of the main functions required in a transmitter. Each block has different levels of flexibility performance demands. Things like forward error correction (FEC) and interleaving operate on bits at the bit rate. These tend to be low-complexity operations in the transmitter, but the respective decoding process at the receiver is a very computationally intensive procedure. The modulator converts bits into symbols in the complex plane for transmission. Again, this is usually straightforward in the transmitter. In the receiver, this process requires computationally expensive frequency, phase, and timing synchronization to properly demodulate digital signals. The pulse shaping filter then bandwidth-limits the signal for transmission.

Before the symbols can be sent to the analog portion of the transmitter through the DAC, we often need some rate conversion step. The DAC and the analog hardware will be clocked at some rate that must be at least twice as fast as the bandwidth of the signal going into the DAC to meet the Nyquist criteria. This requires some rate conversion algorithm that upsamples the signal to match the DAC's sampling rate.

Again, these processes must balance performance with flexibility. For a specified radio standard that uses only one form of modulation or FEC coding, there is no need for flexibility and all of these operations can be realized in hardware. However, for any cognitive radio purpose, we require a great amount of flexibility. It is imperative that the coding and modulation are performed in software to support new and different methods or techniques. On the other hand, given a particular set of analog hardware, the resampling stage is a generic method that operates the same for many different kinds of waveforms, and so it is not required to be as flexible. Furthermore, the sampling rate is directly related to the computational cost and a performance-limiting parameter for software.

An SDR designer must decide which components should be in hardware or software, and what types of processors should run the software based on design needs and trade-offs. Components can be implemented in GPPs, DSPs, FPGAs, and application-specific integrated circuits (ASIC). These elements have been listed in this way to represent the general trend from flexibility to performance with GPPs being highly flexible and versatile while ASICs are designed for a particular purpose but perform their tasks very efficiently. FPGAs are flexible to some extent in that they are reprogrammable, but their programming does not have any real-time dynamics, yet they are more powerful for a given task than a GPP would be.

As we discuss in the following sections, we perform most of the signal processing elements in GPP software. The resampling to match the DAC sampling rate is more universal to waveforms as we mentioned, so we allow the FPGA to take care of this process. The rest is done in analog hardware. The receiver lines are drawn in the respective stages as the transmitter. After the ADC, and FGPA downsamples and filters the signal while the full receive chain is handled in software. Other SDR implementations favor splitting between FPGAs and DSPs with more emphasis on the former. These are design choices to maximize the required behavior.

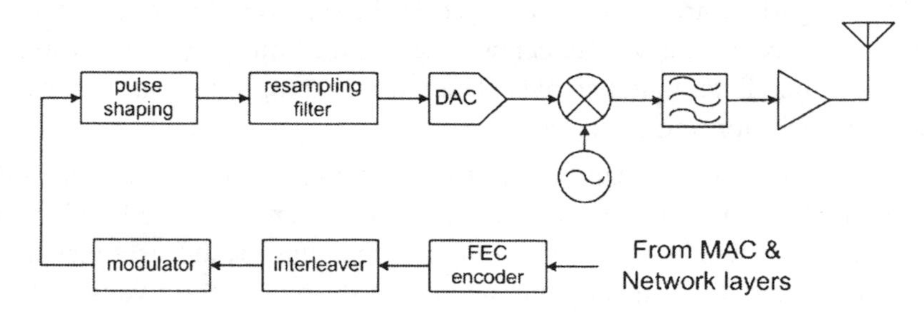

Figure 3.2 High-level SDR physical-layer transmitter with bits received from the upper layers of the protocol stack.

To help us make these decisions, software radio engineers must first understand what systems are capable of performing the different tasks most efficiently. We will first explore what can be done with software before exploring the limitations. From this discussion, we will develop an understanding of what communications capabilities can be expected to run in software and therefore what a cognitive radio can control.

3.2 Benefits of Using SDR

Probably the greatest benefit of SDR technology is the flexibility it can provide. Developing software to perform signal processing offers large opportunities for improving the development cycle. From an operations standpoint, developing and debugging software is much easier, more practical, and more cost-effective than designing and producing hardware like an ASIC where the turnaround time is long and expensive and constitutes a large barrier to entry into the field. From a service provider's perspective, SDR offers easy upgrades and bug fixes in deployed systems. If a new system or waveform is required, as long as there is enough processor power, software updates can be pushed to a system operating in the field [4]. A successful example of this was the recent upgrade of Vanu, Inc.'s mobile base stations that were running a global system for mobile communications (GSM) system and were upgraded to support code division multiple access (CDMA) [5]. This capability saved time and cost of design and deployment, and it lowered the costs to the service provider, who did not have to install a new system. Although in the particular case of Vanu, Inc., the SDR is implemented primarily in general purpose processors, many SDR platforms are being built around FPGAs, which can easily handle software upgrades when changes are infrequent and do not require real-time adaptation of a waveform. While FPGAs offer higher performance for lower power, changes to the FPGA firmware can take on the order of seconds; this is an acceptable lag when updating a system, but not quite fast enough for reconfiguring a waveform in a cognitive radio setting.

Another benefit of software radio is the concept of software reusability. When software is modular and well-written, it can be ported between processors with minimal rewriting required. Unfortunately, this is not entirely the case in today's FPGA-based SDR systems where the software language, generally very high-speed integrated circuit hardware description language (VHDL), is too low level and does not provide sufficient abstraction to be platform independent. We suspect this will change as more influence from the computer science community affects development practices in the SDR world. Nevertheless, in systems that are GPP based, code portability is a major advantage.

Cognitive radio depends on having as much flexibility in waveform design and reconfiguration as possible, and the more flexible the underlying platform, the more useful the cognitive radio is. The case of cognitive radio, unlike over-the-air downloads or service upgrades, requires real-time reconfiguration of much of, if not the entire, waveform. For the reconfiguration, given current technology, a GPP should handle the majority of the signal processing; that is, minimum hardware, maximum GPP.

Another benefit of SDR is that, being software already, it is easy to test individual signal processing blocks, simulate performance, and test behavior in a closed system and then reuse the same software for a real, over-the-air system. Later in this chapter, we discuss the GNU Radio SDR platform used in which we simulate the performance of our cognitive radio prototype. The transmitter and receiver from the simulation are then used for over-the-air experiments by adding the USRP radio front-end instead of a software interconnect.

In software radio, we are benefiting from two parallel development paths, one from the hardware domain and one from software. First, Moore's law is one of our best friends. The technology trend in integrated circuits is constantly giving us more computational resources. Currently, the multicore and parallel processing systems increasingly available on the market are providing us with more processing capabilities at lower power consumption to make SDR an even more promising technology. A second reality of these technology shifts are increasing capabilities within the processing elements themselves. The push of DSP-type elements such as single instruction multiple data (SIMD) into general purpose processors is giving us more throughput at lower cost.

A second factor is the paradigm shift from the early digital emulations of analog circuits to more recent digital building blocks that have no analog counterpart. In his multirate signal processing book, fred harris has shown new ways of thinking about and solving old problems [6]. Exploiting digital signal processing techniques, harris offers many advances that both improve performance and reduce overhead.

Because of the work in both processor technology and signal processing research, the trend is increasingly favoring SDR for implementing wideband, complex, and dynamic radio systems.

3.3 Problems Faced by SDR

Of course, all the benefits of SDR come with a cost such as power consumption, speed, and efficiency. In hardware, the designer can optimize a circuit or chip for a particular purpose that will provide the processing required at the lowest possible power consumption, and hence the name *application specific* integrated circuit. On the other hand, *general purpose* processors provide the flexibility and reuse concepts discussed previously, but they do not achieve the same efficient performance as a hardware system dedicated to a particular waveform.

Many of the problems identified here are engineering challenges that cross a variety of disciplines. Processor technology is stepping up to the computing challenges with multicore techniques, advanced instruction sets like SIMD, and graphics processing units (GPUs) being used in multimedia processing and gaming physics engines [7, 8]. Multicore processors currently offer some of the most incredible advances in general purpose computing power [9], especially with concepts like those used in the IBM CELL processor and future asymmetric multicore processors [10]; that is, different types of cores for different processing purposes. In this type of design, GPP-like cores can provide logic and control while GPU-type cores enable high-speed, efficient signal processing. A further advantage of multicore and multiprocessors systems is that operations can take place in parallel.

Parallelization lends itself directly to SDR processing. First, parallelizing the processing elements allows the receive and transmit paths to operate simultaneously. Second, when segmenting a data stream into blocks of samples, these blocks can be processed simultaneously through different parts of the transmit or receive path. With this structure, the SDR can simultaneously perform different tasks like timing synchronization, demodulation, decoding, and framing.

As these general purpose processing elements are increasing in capabilities and our ability to program them is improving, the cost savings of using these systems instead of ASICs becomes even more compelling. While a general purpose processor can be designed once to solve many different problems, the cost of creating an ASIC mask is getting increasingly expensive. To make it worthwhile, an application requiring an ASIC demands a significant customer base. Applications that can be done with GPPs start making more economic sense because of the reduced hardware cost of entry into a market. Along with employing reusable software, companies can drastically lower cost with standard off-the-shelf processor components and portable software.

There will always be a need for some signal processing such as amplification and high frequency mixing to take place in hardware. Other functions such as multirate processing and filtering between the baseband processing and the ADC and DAC will greatly benefit from implementation in FPGAs. Meanwhile, cognitive radio operates most effectively in general purpose baseband processing for the most flexible systems available. In the next section, GNU Radio implementation provides an example of a GPP-based SDR useful in cognitive radio work.

3.4 GNU Radio Design

One of the most popular SDR implementations is GNU Radio [11], a GNU (the clever recursive acronym for "GNU is Not Unix") project of the Free Software Foundation (FSF) to provide a GPP-based open source software defined radio. We focus on these SDR products extensively in this book for many reasons. First, one of us, Rondeau, is an active developer for the GNU Radio project and so has extensive knowledge of it. Second, GNU Radio is one of the most complete and widely used SDRs for cognitive radio development. GNU Radio and USRPs are used in academic research, with amateur radio enthusiasts, for government needs, and increasingly small business innovations.

As an open source project licensed under the general public license (GPL), these SDR tools make themselves easily and readily available for use in both research and development. With years of work gone into producing the GNU Radio framework and the USRPs, these projects offer a quality foundation for software and cognitive radio work.[1] The free software licensing of GNU Radio also means that anyone reading this book can download and experiment with the ideas we are presenting.

GNU Radio is a software package that provides signal processing *blocks*, discrete components to perform a specific task. Each of these components is a C++ class that a developer can connect to other blocks to create a *flow graph*. A block can be a source with only output ports, a sink with just input ports, or a general block with both inputs and outputs. Currently, the GNU Radio supports many signal processing blocks and a number of waveforms. Blocks include finite impulse response (FIR) and fast Fourier transform (FFT) filters, simple arithmetic operations, complex number processing and transformations, frequency translation, waveform-specific techniques [12], and timing synchronization blocks [13, 14, 15, 16].

In GNU Radio, each separate signal processing block is implemented in C++ and gets built into a library. Python, a high-level programming language (www.python.org) is then used as an interface language to connect the signal processing blocks from the C++ library together. This is performed by "wrapping" the C++ library into a Python module so that Python can call the C++ functions through the "wrapper." Blocks are connected to build a flow graph, an architectural construct that represents the flow of samples through the radio. Blocks can be connected in a hierarchical fashion as well, which allows us to build blocks of more specialized processing capabilities out of smaller, more general blocks. Recently, the project has allowed all of these capabilities to be done in a full C++ environment, so that all GNU Radio functionality is accessible through either a C++ or a Python project. The move to all C++ allows more traditional workspace development of waveforms, applications, and embedded systems.

3.4.1 The Universal Software Radio Peripheral

A software-only SDR does little actual radio communications without a means to get to and from the radio frequency (RF) domain. A device is required to convert between the analog, RF domain and the digital, software world. These devices are referred to as either air interfaces or RF front-ends. A parallel project with the GNU Radio to provide an air interface is the Universal Software Radio Peripheral (USRP) [17]. The USRP is a board that does basic intermediate frequency (IF) processing of up- and downconversion, decimation and interpolation, and filtering. Along with the USRP board are

1. For anyone concerned with learning more about how businesses can use and benefit from open source software, see work by Bruce Perens, http://opensource.org, and the Free Software Foundation at http://www.fsf.org, among others.

a set of daughterboards to perform the final analog up- and downconversion, filtering, and amplification. The USRP provides the air interface to convert between the digital baseband processed in the SDR and the analog, RF domain.

The USRP provides analog to digital sampling with a 12-bit, 64 Msps ADC and digital to analog conversion with a 14-bit, 128 Msps DAC. It can hold two transmitter and two receiver daughterboards at a time. It is controlled and data is transmitted and received over a USB 2.0 interface.

While the USRP and GNU Radio are parallel development projects, they do not necessarily depend on one another as other SDR platforms use the USRP (e.g., [18, 19]) and other RF front ends can use GNU Radio as the signal processing system.

3.4.2 The USRP Version 2

An updated version USRP was recently released, the USRPv2. This design offers a number of improvements to the first generation USRP. First, it uses higher-resolution and faster ADCs and DACs at 14-bits, 100 Msps and 16-bits, 400 Msps, respectively. It only handles one transmitter and one receiver at a time, unlike the USRPv1, but the USRPv2 is equipped with a high-speed serial connector to tie two USRPv2s together for multiboard processing. The USRPv2 includes a more capable FPGA, the Xilinx Spartan 3, than the USRPv1, which uses an Altera Cyclone. The USRPv2 also stores the FPGA image on an SD Card instead of writing it from the host computer when first run as the USRPv1 does. With the SD Card and a MIPs processor image programmed on the Vertex IV, the USRPv2 is capable of complete stand-alone operation. One of the most significant changes between the USRPv1 and USRPv2 is the use of a gigabit Ethernet (GigE) link instead of USB to transfer data. With full-duplex GigE, the USRPv2 can simultaneously transmit and receive 25 MHz bandwidth signals at complex baseband.

Throughout this book, the real differences between the USRPv1 and USRPv2 are unimportant. In the cognitive radio concept, these systems play the same role. We will generically refer to these two systems as the USRP since we will not need to further differentiate their capabilities.

3.4.3 Flow Graphs

The role of GNU Radio ends between the pulse shape filter and the frequency upconversion block in Figure 3.2. Minimal code runs on the USRP; its responsibilities only cover the final stages of filtering, interpolation/decimation, and up-/downconversion. GNU Radio handles the rest of the signal processing to fall in line with the general principle of GNU Radio: flexible, easy to program, and available software for anyone to build SDR waveforms. Therefore, the responsibilities of the USRP only include those parts of the waveform that are, to a great extent, required for all waveforms. Along with this, the USRP also follows open source rules, so all of the code is published and available and anyone is allowed to modify it to perform more processing in the USRP if they so desire.

The flow graph design of the GNU Radio allows for abstraction and visualization. Once instantiated, a signal processing block becomes an object, or node, in the graph. From a graph theory perspective, these graphs are simple, flat graphs with no loops that have at least one source and one sink. Information starts from the source and flows sequentially though the remaining blocks in the graph until it terminates at a sink. Types of sources and sinks include vectors to input or output raw data, files to read to and write from a disk, UDP Ethernet ports to read and write from a network card, generalized signal source blocks for creating sinusoidal signals or noiselike signals, and the USRP to transmit and receive data over this device. A flow graph may include many paths and many sources and sinks so long as there are no loops and there is a connection to each input and output port. Figure 3.3 shows a simple flow graph that reads a signal from a file, filters it, mixes it with a sinusoidal tone, and simultaneously outputs it to a USRP and stores it to another file.

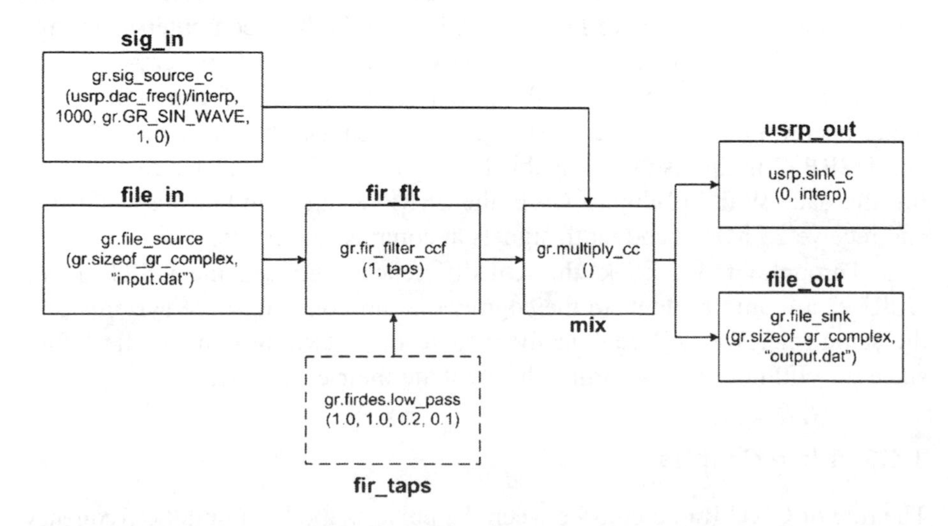

Figure 3.3 Simple GNU Radio flow graph that mixes a filtered signal with a sinusoid, output to a file, and transmitted through the USRP.

We introduce the flow graph concept with this simple application to provide an understanding of how flow graphs can look as well as how to read them. We are using the Python terminology. The blocks in these figures include the name of the GNU Radio block. Most of these are from the *gr* Python module with others coming from the USRP interface module, *usrp*. Any blocks from the *blks2* module are hierarchical blocks of other GNU Radio blocks that are also distributed with GNU Radio. The arguments in parenthesis are the arguments that are passed to the module in the code. The outer name in bold represents a name given to the block.

For this particular graph, we provide in Figure 3.4 the source code that would go along with this graph to realize Figure 3.3. The only new concept to keep in mind in the code is that the *gr.top_block()* is the GNU Radio flow graph object representation.

```
def main():
    # set up the parameters
    interp = 64
    taps = gr.firdes.low_pass(1.0, 1.0, 0.2, 0.1)

    # define the blocks
    sig_in = gr.sig_source_c(usrp.dac_freq()/interp, 1000,
                              gr.GR_SIN_WAVE, 1, 0)
    file_in = gr.file_source(gr.sizeof_gr_complex,
                              "input.dat")
    fir_flt = gr.fir_filter_ccf(1, taps)
    mix = gr.multiply_cc()
    usrp_out = usrp.sink_c(0, interp)
    file_out = gr.file_sink(gr.sizeof_gr_complex,
                             "output.dat")

    # build the flow graph
    tb = gr.top_block()
    tb.connect(file_in, (mix,0)) # file to 1st mixer port
    tb.connect(sig_in, fir_flt)
    tb.connect(fir_flt, (mix,1)) # signal to 2nd mixer port
    tb.connect(mix, usrp_out)
    tb.connect(mix, file_out)
    tb.run() # start the flow graph
```

Figure 3.4 Python source code for the simple GNU Radio flow graph of Figure 3.3.

The graph reads a complex waveform from a file "input.dat" and first filters it. The filter is defined such that the complex envelope of the signal is between the normalized frequencies of ± 0.5 (the Nyquist zone) with a sampling rate of 1. The low-pass filter is designed with respect to the normalized frequency with a bandwidth of 0.2 and a transition width of 0.1. The filtered signal is mixed with a 1-kHz complex sinusoid of ± 1 V and 0 V DC offset. The mixed signal is then stored in a file "output.dat" and passed to a USRP sink to transmit over the air. The USRP takes the parameter *interp* to set the interpolation rate required to match the 128-Msps DAC. Not shown are some of the other commands required to set up the USRP properly to transmit on a particular frequency.

Another property of the flow graph blocks is that developers can build hierarchical blocks that encompass many lower-level blocks. This capability enhances the levels of abstraction in the software. For example, the current GNU Radio distribution includes a differential binary phase shift keying (DBPSK) modulator as a hierarchical block of blocks that perform tasks like symbol mapping, differential encoding, and root raised cosine (RRC) pulse shaping.

3.4.4 Parallel Programming in GNU Radio

Another advance in GNU Radio is to take advantage of the multicore and other parallel processing systems that have become the mainstream in processors. The original flow graph scheduler in GNU Radio was a single-thread per flow graph. Blocks in the flow graph were flattened into a one-dimension graph from sources to sinks and each block was sequentially run.

GNU Radio blocks operate on chunks of data to where large blocks are passed around in memory. This is important to reduce memory and bus access overhead. Under this condition, it is easy to see how multiple blocks could run simultaneously in a flow graph. While one block is processing its current chunk of data, the previous block in the graph can be processing the next chunk of data.

A multithreaded scheduler is now available in GNU Radio, currently implemented as a thread-per-block model, where each block runs in its own thread as its name might suggest. When the processing resources are available, each thread is loaded into its own processor core. In tests, Blossom [20] shows a near-linear improvement in the speed of processing. He also points out in this paper the need to balance the load among the threads. There is overhead in context-switching if running multiple threads on a processor as well as passing data along a bus between processors. Each block must then process enough data to overcome this overhead. Threads should also take about the same amount of time to process their chunks of data so that they are not waiting on other blocks in the system to finish. This is not currently considered in the thread-per-block model because each block may have different computational requirements. Still, we see a significant improvement in processing capabilities of GNU Radio because of this change. It is also leading to the next concept of multithreaded scheduling to group and schedule blocks among threads that do better load balancing.

The improvements that we are seeing from the multithreaded models of GNU Radio indicate a positive trend for SDR. As the processors become more parallel, as is the obvious trend from companies like Intel, IBM, and AMD, SDR becomes more powerful. As Blossom's numbers suggest, for each extra core in a processor, we get an almost equal improvement in speed.

3.4.5 Flow Graph for Simulation and Experimentation

One of the benefits of a GPP-based SDR is the easy transition between testing and operation. In this section, we describe the flow graph for the transmitter and receiver paths of the SDR used in the experiments of Chapter 8. The design and implementation is such that the endpoints of either chain may connect to any source for the receiver or sink for the transmitter. As such, we have a system where we can run the transmitter into a channel model and then directly into the receiver chain. With this setup, we can test the properties of the system in a known environment. It is then trivial to replace the channel with a USRP sink or source so the transmitter sends the same signals over the air to be received by another USRP running the receiver flow graph.

The simulation design is shown in the following figures. Figure 3.5 gives the big picture of the simulation and is made up of many GNU Radio blocks as well as some hierarchical blocks, represented as shaded blocks. The simulation generates a signal using *txpath*. The bandwidth of the overall simulation is set in the next two resampling blocks. The process of interpolating adds samples to a digital signal, which increases the sampling rate by the interpolation factor. This process has the effect of increasing the bandwidth of the overall system. The *tx_resample* block performs both interpolation and decimation to allow fractional changes in the symbol rate and first sets the bandwidth of the signal based on the waveform parameter, and *interp* sets the overall bandwidth of the system, providing a "spectrum" for the simulated transceiver and interferers to share and interact within. The *tx_mix* block upconverts the transmitted signal to some frequency set by the numerically controlled oscillator (NCO) *tx_nco* within the system bandwidth. The upconverted signal goes through a channel model described in Figure 3.6 with added interference signals at some bandwidth and center frequency of their own. The signal is then passed to the receiver chain. The receiver first downconverts the signal from the center frequency back to baseband, goes through a channel filter that resamples in reverse of *tx_resample* and

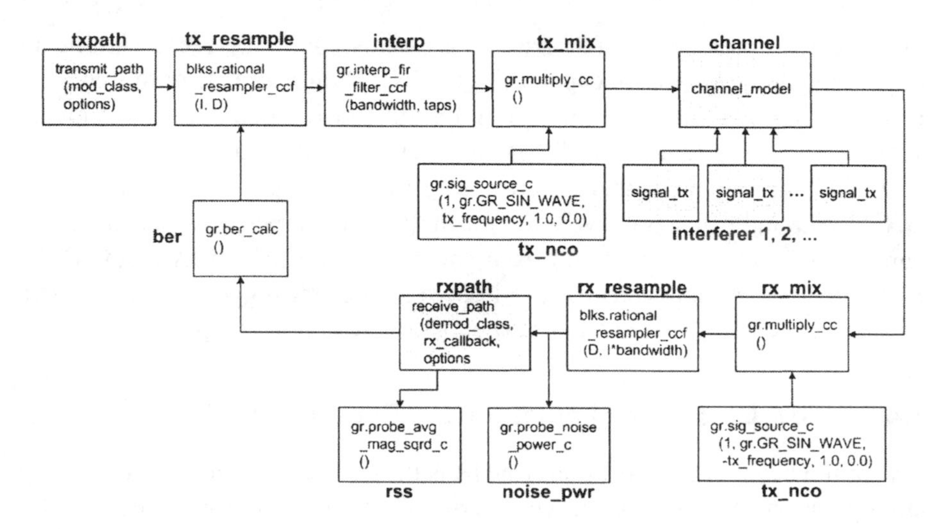

Figure 3.5 Radio simulation model. White boxes indicate low-level blocks while shaded blocks indicate hierarchical blocks.

interp. After filtering and resampling, the resulting signal is the transmitted signal plus noise plus any interference that exists within the signal bandwidth. The received signal then goes into *rxpath* where it is demodulated. Out of the channel filter, *noise_pwr* calculates the noise power of the channel. The block *rss* calculates the received signal strength. Both of these calculations are described below, and together, they are used to calculate the SNR. Finally, the system calculates the BER by taking in the transmitted bits and compares them to the bits received after passing through the channel and demodulator. The first one thousand bits of the input are dropped to ignore any transients in the system.

The channel model of Figure 3.6 adds Gaussian white noise at a given noise voltage calculated from the simulation's noise floor, and a frequency offset used to represent frequency differences between the transmitter and receiver. The path loss is simply modeled as a multiplier with a constant value calculated externally that represents the distance, given a particular path loss model. The channel model is easily extended to include multipath by using a FIR filter with a given set of taps. All of the values that the channel model uses are passed externally to allow representation of different mathematical models.

Figure 3.7 shows the flow graph used to create interference signals. A signal is modulated with any digital modulator from the GNU Radio blocks, given a specific amplitude for a comparative power, interpolated to give it a

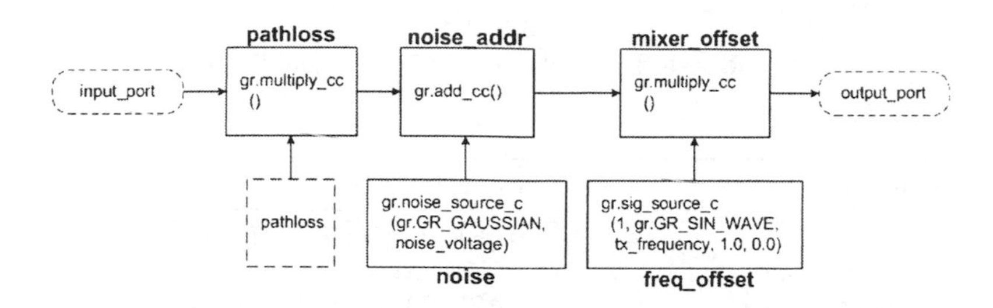

Figure 3.6 Flow graph of simulated channel model with pathloss power loss, additive noise, and frequency offset.

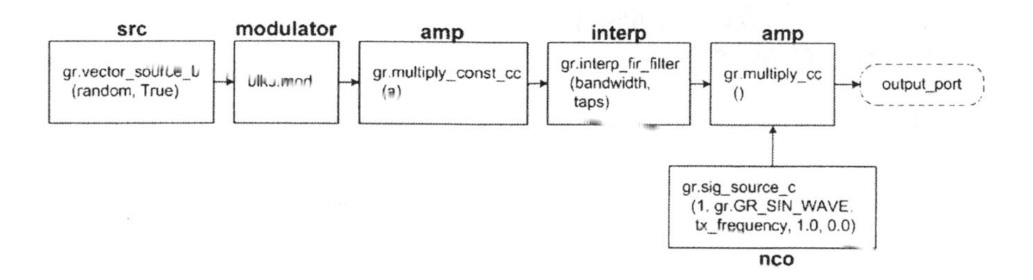

Figure 3.7 Flow graph of simulated interferer.

specific bandwidth, and mixed with a normalized frequency offset to place it properly in the simulation's spectrum.

The final hierarchical blocks are the transmit and receive paths. They are not shown in figures here due to their simplicity. All they do is wrap a modulator or demodulator into a block that has accessor functions that read and write data from a higher layer while the modulators, hierarchical blocks themselves, are part of the standard distribution of GNU Radio.

3.4.6 Available Knobs and Meters

The knobs of Table 3.1 are available to the simulation.

The meters are calculated as follows. The system estimates the noise power during dead-time on the channel. If the received signal strength is less than some specified threshold, the system assumes there are no transmitting radios and can therefore calculate the noise power in the channel. The noise power is the variance of the samples calculated using Knuth's online (that is, it can be calculated on streaming data) algorithm [21]. The listing (in Python) of the algorithm is provided here. Of course, the variance calculation collapses

Table 3.1
Knobs and Meters Available to the GNU Radio Simulation

Knob Name	Knob Settings
Modulation	(D)BPSK, (D)QPSK, (D)8PSK, GMSK
Transmit power	0 – 20 (dBm)
Symbol rate	0.125, 0.25, 0.5, 1 (normalized sps)
Shaping filter roll-off	0.1 – 1.0, steps of 0.01
Normalized frequency	-1.0 – 1.0, steps of 0.01
Frame size	100 – 1500, steps of 1
Meters	
Received signal strength (dBm)	
Noise power (dBm)	
Bit error rate	
Packet error rate	
Path loss	
Signal-to-noise ratio	

to the average magnitude squared calculation when the mean of the signal is 0.

```
def variance(data):
    count = 0
    mean = 0
    sum = 0
    for x in data:
        count += 1
        delta = x - mean
        mean = mean + delta/count
        sum += delta*(x - mean)
    variance = sum/(count - 1)
```

In the GNU Radio block for calculating the variance, the number of items available for processing is a variable passed to the function. The loop is calculated over all of the available items, then the variance is calculated before the block exits.

The received signal strength is first calculated by taking the average magnitude squared of the samples when the transmitter is known to be running. However, this estimation is biased by the noise power, especially at low SNRs, so the final signal strength estimation is shown in (3.1), which, by subtracting the noise power, provides a good estimation of the signal strength.

$$P = 10\log_{10}\left(\frac{1}{K}\sum_{i=0}^{K-1}|x_i|^2 - n\right) \quad \text{dBm} \tag{3.1}$$

Here the x_is are the time samples in millivolts and n is the estimated noise power in milliwatts. The GNU Radio block that implements the average magnitude squared simply calculates the norm of each complex signal and uses a single-pole infinite impulse response (IIR) filter to average the results.

Path loss and the energy per bit to noise ratio (E_b/N_0) are easy to estimate since the radios know the transmitted power and received power, the difference is therefore the path loss, and the radio has already estimated both the received power and noise power. Both of these values are then converted into the energy per bit (E_b) and noise density (N_0) to more accurately reflect the modulation behavior. The simulation calculates E_b from the signal power by dividing by the bit rate, $R_b = kR_s$ where k is the number of bits per symbol and R_s is the symbol rate. The calculation of the noise energy comes from dividing the received power by the receiver bandwidth. In this case, the simulation is all in the digital domain, so the noise power reflects the amount of power in the symbol sampling interval, which is represented by the number of samples per symbol. The noise energy then comes from calculating $N_0 = N/sps$, the noise power divided by the number of samples per symbol.

In our simulation, the BER calculation is a more complicated meter. Figure 3.5 shows a BER calculation block (gr_ber_calc) with an input from the transmit path and one from the receive path. The BER is not simply a comparison of the transmitted and received bits as a one-to-one comparison because of the delay introduced by the blocks in the flow graph. The received bits are therefore delayed some amount from the transmitted bits. Furthermore, we observed that the delay is different for different types of modulations, changes with the symbol rate, and, for some types of modulations, the delay is not consistent between runs (GMSK, for instance, would have a delay of 7 or 8 samples during any given run). Consequently, the BER calculation block calculates and compensates for the delay. To do this, it waits to see a particular start code from both the transmitter and receiver and calculates the delay between corresponding samples. To avoid the problem of losing the start code in the noise, the start code packet is always transmitted with a much higher SNR (near infinite in the simulation by setting the noise voltage to 0).

The BER estimation works well in the simulation, though the over-the-air procedure has not been developed. For performing the analysis over the air, more synchronization is required between the transmitter and receiver and a procedure is required to know the actual transmitted bits for proper comparison, which is usually done using a pseudorandom sequence generator with known seeds. The transmitted bits could be known at the beginning, but synchronization is required if a packet is lost. The bits could be calculated using a seed passed in the packet, which could be corrupted itself, or a secondary transmit path, such as over Ethernet, could be used with guaranteed fidelity, although this still requires synchronization. When a packet is lost, it is due to either corruption in the access code or the header, and the BER calculation needs to account for the lost bits without assuming all the bits in the packet were lost.

There are many ways to solve the problem of over-the-air BER testing, and many systems implement BER tests, but the problem has not been solved in GNU Radio. Because we can get the performance analysis in the simulation, though, we do not solve this problem here and rely on packet errors as a performance measure during our over-the-air experiments. The packet error rate is an easy calculation because the packet numbers are included with the packets. The PER is calculated by knowing the number of packets transmitted from the packet number (or by being told) as well as the number of good packets received, which is determined by a 32-bit cyclic redundancy check (CRC).

Finally, the interference power is measured using a simple energy detector. The receiver opens up its bandwidth to sample the entire band of interest, and, knowing the noise power from above, finds the frequencies where the signals exceed a certain threshold (e.g., 5 or 10 dB) above the noise floor and where the signals fall below the threshold. These frequencies are stored along with the average power between them to build an interference "map" of the channel. The cognitive radio can then use the map to understand the interference potential in a given bandwidth. Chapter 2 discussed the use of these meters in the cognitive engine and presented the XML format used to represent the information.

3.5 Conclusions

This chapter provides both a basic review of software radio technology and introduces the SDR system used later in the experiments. The GNU Radio platform solves a number of problems in working with SDR. First, it provides an affordable platform (free software plus inexpensive hardware) that is open and therefore transparent to analysis as well as upgrades.

One of the authors, Rondeau, wrote large portions of the digital communications capabilities in the current GNU Radio distribution. These were developed because of the need to perform the experiments for this work but which also led to contributions to the SDR community. He is still actively involved in GNU Radio development.

From a research and scientific point of view, the publicly available revision logs of the software and the open software platform enable the use of the scientific method for SDR and CR experiments. Anyone can use the same software platform to test or compare results.

With the available GNU Radio platform and a description of the knobs and meters, the next chapter introduces the concept of waveform optimization that uses the information from the meters to make decisions on how to tune the knobs.

References

[1] J. H. Reed, *Software Radio: A Modern Approach to Radio Engineering*, Upper Saddle River, NJ: Prentice Hall, 2002.

[2] W. Tuttlebee, *Software Defined Radio: Enabling Technologies*, New York: John Wiley & Sons, 2002.

[3] B. Le, T. W. Rondeau, J. H. Reed, and C. W. Bostian, "Analog-to-Digital Converters: A Review of the Past, Present, and Future," *IEEE Sig. Proc. Mag.*, Vol. 22, No. 6, pp. 69 – 77, Nov. 2005.

[4] M. Dillinger and R. Becher, "Decentralized Software Distribution for SDR Terminals," *IEEE Trans. Wireless Communications*, Vol. 9, pp. 20 – 25, 2002.

[5] J. Kumagai, "Radio Revolutionaries," *IEEE Spectrum*, Vol. 44, No. 1, pp. 28 – 32, Jan. 2007.

[6] f. harris, *Multirate Signal Processing for Communication Systems*. Upper Saddle River, NJ: Prentice Hall, 2004.

[7] General Purpose Graphical Processing Units (GPGPU), [URL: http://www.gpgpu.org/], 2009.

[8] Open Computing Language (OpenCL), [URL: http://www.khronos.org/opencl/), 2009.

[9] K. Asanovic, R. Bodik, B. C. Catanzaro, J. J. Gebis, P. Husbands, K. Keutzer, D. A. Patterson, W. L. Plishker, J. Shalf, S. W. Williams, and K. A. Yelick, "The Landscape of Parallel Computing Research: A View from Berkeley," Technical Report UCB/EECS-2006-183, Berkeley, Dec. 2006.

[10] IBM CELL Research, [URL: http://www.research.ibm.com/cell/], 2007.

[11] Free Software Foundation. GNU Radio, [URL: http://gnuradio.org/trac], 2009.

[12] T. W. Rondeau, M. Ettus, and R. W. McGwier, "Open Source Transparency for OFDM Experimentation," *Software Defined Radio Forum Technical Conference*, 2008.

[13] J. P. Costas, "Synchronous communications," *Proceedings of the IRE*, Vol. 44, pp. 1713 – 1718, 1956.

[14] J. Feigin, "Practical Costas Loop Design: Designing a Simple and Inexpensive BPSK Costas Loop Carrier Recovery Circuit," *RF Signal Processing*, pp. 20 – 36, 2002.

[15] K. Mueller and M. Müller, "Timing Recovery in Digital Synchronous Data Receivers," *IEEE Trans. Communications*, Vol. 24, No. 5, pp. 516 – 531, 1976.

[16] G. R. Danesfahani and T.G. Jeans, "Optimisation of Modified Mueller and Muller Algorithm," *Electronics Letters*, Vol. 31, No. 13, pp. 1032 – 1033, 1995.

[17] Ettus Research, LLC, Universal Software Radio Peripheral, [URL: http://ettus.com], 2009.

[18] P. MacKenzie, *SOftware and Reconfigurability for Software Radio Systems*, Ph.D. diss., Trinity College Dublin, Ireland, 2004.

[19] M. Robert, S. Sayed, C. Aguayo, R. Menon, K. Channak, C. V. Valk, C. Neely, T. Tsou, J. Mandeville, and J. H. Reed, "OSSIE: Open Source SCA for Researchers," *Software Defined Radio Forum Technical Conference*, 2004.

[20] E. Blossom, "gcell - An SPE Scheduler and Asynchronous RPC Mechanism for the Cell Broadband Engine," *Software Defined Radio Forum Technical Conference*, 2008.

[21] D. E. Knuth, *The Art of Computer Programming, Volume 2: Seminumerical Algorithms*, Vol. 3, Boston: Addison-Wesley, 1998.

4

Optimization of Radio Resources

As shown in the previous chapter, a cognitive radio uses AI to adapt and optimize the performance of a radio platform, specifically an SDR. In this chapter, we explain the concepts of radio resource optimization with a particular interest in understanding the physical layer in terms of a set of objective functions. This discussion establishes a way of thinking about how to optimize certain objectives. Although this chapter includes only a subset of the total possible objectives along with some basic approximations, the development should provide the fundamentals that will enable further extension and enhancement as more information and capabilities arise.

4.1 Objective Space

As in any problem, we first have to define the vocabulary. In optimization problems, this vocabulary is the objective space, which is the set of all possible solutions to a problem, often over a multidimensional set of objective functions. The cognitive radio objective space describes the radio resources that determine radio behavior.

A radio consumes resources while communicating, thereby depriving other radios access to those same resources. Spectrum is the key communications resource. Economically speaking, spectrum is a reusable resource because after one radio is done *consuming* the spectrum, it is still available for other radios to use. In fact, radios can share spectrum in space, time, transmit power, and other methods that allow signals to coexist on the same frequencies. Spectrum sharing and reuse is accomplished through numerous techniques such as spatial distribution as in cellular infrastructures or beam-forming antennas [1]. Standard practices like time division multiple access (TDMA) and frequency division multiple access (FDMA) allow sharing and reuse in time and frequency domains, respectively. DSA technology is developing to provide intelligent schemes that use spectrum during times when its primary occupants are silent [2]. Concepts such as an ultra-wideband underlay and interference temperature are all methods that manage transmit power to allow coexistence with other radio systems [3, 4]. Finally, transmitting orthogonal signals like direct sequence spread spectrum (DSSS) reduce interference between signals occupying same spectrum at the same time. The task is then to properly use the resources to provide appropriate sharing among all radios while maintaining the proper level of QoS.

Each user has a different and subjective perspective on quality of service based on the radio's performance. A user may require high data rates, low latency, or long battery life depending on the situation for which he or she is using the radio service. Video conferencing requires high data rates and low latency, while voice calls require low latency but have significantly relaxed requirements for average throughput. On the other hand, checking stock prices or even email has low requirements for speed. Long journeys without access to a power source would increase a preference for low power consumption and longer up-time.

Each node in a network can look at resource allocation as an optimization problem with two potential goals. First, it can attempt to optimize its own ability to communicate by maximizing its use of resources; this would be called a *greedy* approach. The other way is to look at resource utilization from a needs perspective; that is, resources are sought only to support the needs of the service. Using more resources is wasteful while using less reduces the quality of service. Resource allocation on either side of what is required is inefficient. Of course, there is a third way of looking at resource allocation, and this is to look at it from a global perspective where the utilization of resources by all nodes is taken into account [5]. While there is significant benefit to this approach, the argument we present here comes down to the personal perspectives on the quality of service offered, which is therefore to see how well the radios use the resources to provide the QoS desired by the user. By maintaining the balance between the use of resources to provide the QoS as well as avoiding overuse of resources, a cognitive radio provides users with proper service while minimizing resource consumption and so allowing others their share of the resources.

Another way to look at this is to consider the objective analysis of resources. If a service requires high data rates, low bit error rates, and low latency yet keeps an eye on the power consumed by the radio while using a particular waveform, the radio has the potential to balance these requirements and design a waveform that properly meets all of the objectives. High transmit power and high-order modulation and coding schemes may provide high throughput, but the power to transmit the signal and the power to receive the signal make the waveform inefficient. The waveform has not met the power consumption objective. On the other hand, another modulation, coding, and frequency could provide a slightly lower data rate but with much better power performance. Because of the uneven trade-off, the second waveform wins out as a better use of resources while still achieving the required QoS.

Unfortunately, simple BER or SINR calculations do not tell the entire picture of the waveform and the QoS. Bit error rate in a voice system does not necessarily relate to the quality of service if a poor vocoder (voice coding/decoding) is used or the propagation path has high burst errors. Many factors impact the resulting QoS of wireless communications systems, and so joint optimization and analysis are required.

From this discussion, it follows that the optimization of radio resource allocation is a multiobjective problem: the analysis of multiple objectives on the decision-making process (also know as a many-objective problem). The next section discusses the concept of multiobjective optimization and the objectives used in the optimization of a radio's PHY layer.

4.2 Multiobjective Optimization: Objective Functions

Multiobjective optimization has a long history in mathematics, operations research, and economics. A relatively old book by Hwang and Syeed [6] provides a comprehensive mathematical introduction to the study of multiobjective optimization. We introduced this concept for wireless communications in [7] and reproduce and extend that discussion here. Zitzler [8] gives an overview of multiobjective problems and presents the basic formula for defining a multiobjective decision making (MODM) problem as shown in (4.1).

$$\begin{aligned} min/max\ \bar{y} &= f(\bar{x}) = [f_1(\bar{x}), ..., f_n(\bar{x})] \\ \text{subject to: } \bar{x} &= (x_1, x_2, ..., x_m) \in X \\ \bar{y} &= (y_1, y_2, ..., y_n) \in Y \end{aligned} \tag{4.1}$$

This equation defines n dimensions in the search space where each objective function $f_n(\bar{x})$ evaluates the nth objective. The set $\bar{x}$ defines the set of input parameters that the algorithm has control over, and $\bar{y}$ is the set of objectives computed by the objective functions. Both of these may be constrained to some space, X and Y, depending on real-world considerations like available radio resources ($\bar{y}$) or radio capabilities ($\bar{x}$). The solutions to multiobjective problems lie on the *Pareto front*, which is the set of input parameters, $\bar{x}$, that defines the nondominated solutions, $\bar{y}$, in any dimension. A key factor in multiobjective problems is that many, if not all, objectives compete for dominance. For example, it is impossible to both minimize BER and minimize transmit power. Multiobjective optimizations often consist of such trade-offs in goals when finding the best solutions available [8, 9]. In the above example, the cognitive radio has a trade-off space where it wants to maximize the throughput but must minimize the power consumption. Each of the extremes lies on the Pareto front as does a set of solutions that offers compromises between the extremes. These are more likely the solutions a cognitive radio is interested in as it maximizes the QoS and minimizes resource consumption.

The following sections describe the different objectives currently identified and defined. For each objective, we list the required knobs, meters, and other objective functions required to analyze the function. When an objective depends on another objective, the knobs of meters used by the dependent objective are not listed. As an example, all BER calculations depend on the signal-to-noise ratio where the signal power is a function of the transmitter power (a knob) and path loss (a meter). The system noise floor is a meter and the noise power in the channel depends on the channel bandwidth, which is another objective function. Since the cognitive engine can change the bandwidth by adjusting the filters and the symbol rate, neither of these are listed as direct dependencies of the BER objective because they are already linked due to the bandwidth as an objective dependency.

Except when otherwise noted, we took most of these functions straight out of a standard communications text such as Proakis or Couch [10, 11]. As we pursue this discussion, we will be talking about the specific capabilities and limitations of the VT-CWT cognitive engine we have built and not general limitations of cognitive engines.

4.2.1 Bit Error Rate (BER)

Dependencies

Knobs: transmitter power, modulation type

Meters: noise power, channel type, path loss

Objectives: bandwidth

Definitions

γ = energy per bit to noise energy ration (E_b/N_0)
P_T = transmit power (effective isotropic radiated power (EIRP)) (dBm)
L = estimated path loss (dB)

B = bandwidth calculated as in Section 4.2.2 (Hz)
M = number of symbols in the modulation's alphabet
R_s = symbol rate (sps)
N_0 = noise floor (J)

Bit error rate (BER) is an important objective for all digital communications' needs. It provides a baseline for the amount of information transferred, and so understanding it in light of the design of a waveform under certain channel conditions is therefore necessary. Unfortunately, BER calculations depend heavily on the type of channel and type of modulation, and so the cognitive engine must know the formula for each modulation type the radio is capable of using and the channel types it is likely to see during operation. In this section, we present a number of bit error rate formulas that have been programmed into the cognitive engine.

To predict the BER of a waveform under a given set of conditions, the cognitive engine requires knowledge of certain environmental conditions. All BER calculations depend fundamentally on the signal-to-noise ratio (SNR) at the receiver. When calculating the BER, however, the cognitive engine does so knowing the transmitter power as a knob that it can set. The cognitive engine therefore needs an estimate of the noise power at the receiver and received power due to path loss. Furthermore, since the noise power changes with the channel bandwidth, the cognitive radio must estimate the noise floor for use in the calculation of the noise power given the bandwidth. With this knowledge, the cognitive engine, knowing the transmitted power, the path loss, and the noise floor, can estimate the received power and noise power for the BER calculation. The cognitive engine will also require an understanding of the type of channel. For the purposes of this work, we focus on additive white Gaussian noise (AWGN) channels, though we provide formulas to calculate BER in Rayleigh, Ricean, and Nakagami-m fading channels as well in Appendix B. This decision to use AWGN channels was made largely because we have do not have a sensor available in the cognitive engine that can determine the channel type, though there are methods of doing this as well as using channel impulse responses for the estimation of BER [12]. AWGN channels also provide more mathematically simple channels to perform simulations and calculations, and so it provides us with an understanding of the optimization behavior in a well-known channel.

Another missing part of these BER calculations is the effect of an interferer. It is not a simple matter of using SINR instead of SNR in the BER calculations since the BER equations depend on the noise power having a Gaussian distribution while interference power does not. The inclusion of interference power is complicated and depends on the types of signals in use (except in such cases as CDMA [13]). Currently, the cognitive engine does not have the capabilities to make this kind of distinction in its optimization process; however, significant work has gone into developing signal and modulation classification schemes since [14] up to [15], and the addition of this information could lead to a better understanding of the BER due to SINR.

As a quick side note, while the normal representations of BER formulas tend to use the Q-function, all of the equations listed here use the complementary error function ($erfc$). We use this because of a few simple approximations of $erfc$ from [16, 17]. To review:

$$Q(x) = \frac{1}{2} erfc\left(\frac{x}{\sqrt{2}}\right) = \frac{1}{\sqrt{2\pi}} \int_x^{\infty} \exp\left(-t^2/2\right) dt \tag{4.2}$$

For $x \geq 3$:

$$erfc(x) = \frac{\exp\left(-x^2\right)}{x\sqrt{\pi}} \left(1 - \frac{1}{2x^2} + \frac{(1)(3)}{2^2 x^4} - \frac{(1)(3)(5)}{2^3 x^6} + \ldots\right) \tag{4.3}$$

For $x < 3$:

$$erfc(x) = 1 - \frac{2}{\sqrt{\pi}} \sum_{n=0}^{\infty} \frac{(-1)^n x^{2n+1}}{n!(2n+1)} \tag{4.4}$$

Each of these equations is easily coded into a simple algorithm to iterate over a finite number of terms in the series. A very close approximation can be achieved with 10 terms for (4.3) and for 40 terms in (4.4). Figure 4.1 shows the exact plot of the $erfc$ function compared to the estimate for $0 \leq x \leq 5$ along with the percent error. There is a small spike in the estimation at $x = 3$ of 0.01% due to a discontinuity at the point where one estimation ends and the other picks up. This can be adjusted by changing where the formulas hand off and by increasing the number of terms in the series for (4.4). Decker addresses the problem of using his approximation for values smaller than three in his paper [16]. So although (4.4) is capable of representing the complementary error function for increasingly large values of x, the number of terms in the series must grow as well. Equation (4.3) represents the $erfc$ for large values of x in a simpler series formula for a more efficient calculation.

We bring up this discussion to point out considerations that have to go into the definition of the objective functions. Accuracy in the modeling is only one consideration while issues of computation time can impact the overall effectiveness of the objective in the optimization process.

The signal power is calculated from the waveform's transmit power and the estimated path loss in the channel. The noise power in the channel

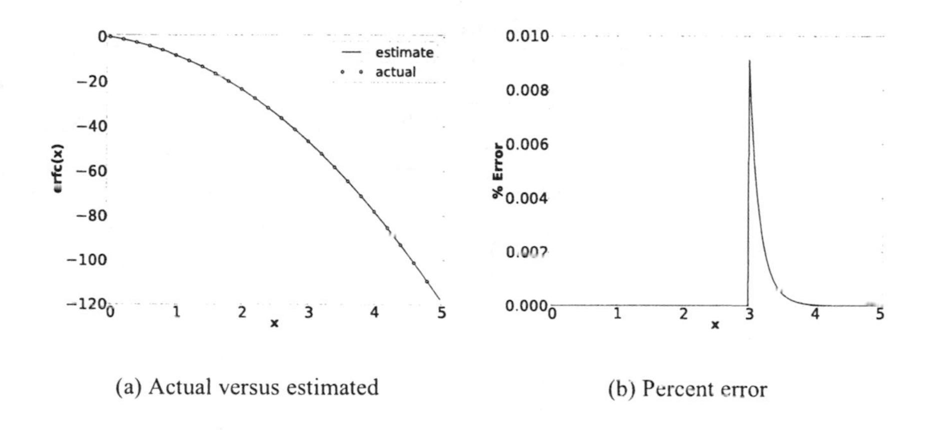

(a) Actual versus estimated

(b) Percent error

Figure 4.1 Comparison of the exact $erfc$ function to the estimation.

is calculated from the bandwidth of the waveform and the noise floor. The BER formulas are presented using E_b/N_0, or the ratio of the energy per bit to the noise power spectral density, as the standard representation of the signal quality. If S is the signal power, then the energy per symbol is S/R_s, since the symbol rate is the inverse of the symbol time. The energy per bit is the energy per symbol divided by the number of bits per symbol, $\log_2(M)$. The waveform sets the transmitted signal power, so the received signal strength is approximated by (4.5), where P_T is the transmitted power and L is the path loss.

$$S = P_T - L \qquad \text{(dBm)} \tag{4.5}$$

The noise power spectral density is the noise per Hz, calculated in (4.6), where k_B is Boltzmann's constant (1.38×10^{-23} J/K) and T is the system noise temperature (K). The noise power in the channel is the noise power spectral density multiplied by the channel bandwidth, B.

$$\begin{aligned} N_0 &= k_B T \qquad \text{(J)} \\ N &= 10\log_{10}(BN_0) + 30 \qquad \text{(dBm)} \end{aligned} \tag{4.6}$$

The E_b/N_0 is a measurement independent of the signal bandwidth as given in (4.7), where $S - N$ is the SNR.

$$E_b/N_0 = \gamma = 10\log_{10}\left(\frac{B}{R_s \log_2(M)}\right) + (S - N) \qquad \text{(dB)} \tag{4.7}$$

Table 4.1
GMSK BER Correction Factor for (4.8)

BT	Correction factor (α)
0.1	0.25
0.2	0.60
0.3	0.77
0.4	0.90
0.5	0.95
0.6	0.97
0.7	0.98
0.8	0.99

Each of the equations below provides the formula for BER calculation of a particular modulation given an AWGN channel.

GMSK (noncoherent demodulator):

$$P_e = \frac{1}{2} \exp(-\alpha\gamma) \tag{4.8}$$

Where α is an adjustment factor to the energy based on the bandwidth-time product (BT) factor of the Gaussian filter [18]. Table 4.1 provides estimates of α over some values of BT based on the work of Murota and Hirade [19].

BPSK:

$$P_e = \frac{1}{2} erfc(\sqrt{\gamma}) \tag{4.9}$$

M-PSK ($M > 2$):

$$P_e = \frac{1}{\log_2(M)} erfc\left(\sin\left(\frac{\pi}{M}\right)\sqrt{\log_2 M}\sqrt{\gamma}\right) \tag{4.10}$$

DBPSK:

$$P_e = \frac{1}{2} \exp(-\gamma) \tag{4.11}$$

DQPSK:

$$P_e = Q_M(a,b) - \frac{1}{2} I_0(ab) \exp\left(-\frac{a^2+b^2}{2}\right)$$
$$a = \sqrt{2\gamma\left(1-\sqrt{\frac{1}{2}}\right)} \qquad (4.12)$$
$$b = \sqrt{2\gamma\left(1+\sqrt{\frac{1}{2}}\right)}$$

Where $Q_M(a,b)$ is the Marcum Q-function and $I_0(ab)$ is the modified Bessel function of the first kind and order zero [10].

Unfortunately, there is no formula for the performance of D8PSK. In the objective analysis of this modulation, we assume it will perform like 8PSK with a 2-dB loss in the SNR.

4.2.2 Bandwidth (Hz)

Dependencies

Knobs: modulation type, symbol rate, pulse shape filter

Meters: none

Objectives: none

Definitions

k = number of bits per symbol

R_s = symbol rate (sps)

r = property of the pulse shape filter (roll-off factor in RRC or bandwidth-time product in a Gaussian filter)

Bandwidth is an objective that is also used in many other objective calculations. It appears in the bit error rate, interference power, spectral efficiency, and throughput. It is a direct measurement of how much spectrum is occupied by the radios. This objective offers a good indication of the spectral resources occupied and therefore measures the trade-offs we have been discussing.

In calculating the bandwidth, instead of using the raised cosine (RC) Nyquist pulse shaping, an RRC filter in both the transmitter and receiver provides a more practical implementation. Over the air, the pulse is shaped by a single RRC filter while the second RRC filter in the receiver shapes the received signal as though it was passed through a single raised cosine filter to reduce the intersymbol interference (ISI). Many narrowband digital modulations use RC pulses, including the M-PSK waveforms used in this work. For these, the approximate null-to-null bandwidth is calculated in (4.13). In this equation, the roll-off factor of the RRC filter is defined as r.

$$\begin{aligned} &B = \frac{R_s}{2} K (1 + r) \qquad \text{(Hz)} \\ &\text{where } K \text{ is the number of frequency dimensions} \\ &K = 1 \text{ for } M\text{-PSK and } M\text{-QAM signals} \\ &K = M \text{ for } M\text{-FSK} \end{aligned} \tag{4.13}$$

With Gaussian-shaped filters, specifically in GMSK, r represents the BT product defined for the 3-dB cutoff frequency of the Gaussian filter. Admittedly, taking this as a formula for the bandwidth is not a fair comparison to the null-to-null formula used for the RRC filters. Unfortunately, though, there is no good closed-form solution to make the same bandwidth comparison. Since the cognitive engine is designed for flexibility in the objective functions, it is trivial to fix this later with a new, more accurate objective function.

$$B = rR_s \qquad \text{(Hz)} \tag{4.14}$$

4.2.3 Spectral Efficiency (bits/Hz)

Dependencies

Knobs: modulation type, symbol rate

Meters: none

Objectives: bandwidth

Definitions

k = number of bits per symbol
R_s = symbol rate (sps)
B = bandwidth calculated as in Section 4.2.2 (Hz)

Spectral efficiency represents the amount of information transferred in a given channel and is measured in bits per second per Hertz (bps/Hz). Although this concept is directly related to both bandwidth and throughput, we develop it as a separate objective in order to represent the quality of service needs more thoroughly. When choosing to measure spectral efficiency, it offers another dimension to determine how suited the waveform is to a particular need when judged by both bandwidth occupancy and data rates. Minimizing bandwidth would push the optimization to use a small symbol rate and a conservative bandwidth modulation like GMSK, while maximizing throughput would push for high symbol rates and high-order modulations. Spectral efficiency helps shape the decision space by biasing the solution towards a symbol rate and modulation type that provides high bandwidth efficiency and produces better data rates for given spectrum availability.

$$\eta = \frac{R_s k}{B} \qquad \text{(bps/Hz)} \tag{4.15}$$

4.2.4 Interference

Dependencies

Knobs: frequency

Meters: interference map

Objectives: bandwidth

Definitions

f_c = center frequency of waveform (Hz)

$I(f)$ = interference spectral power density at frequency f from interference map (mW/Hz)

B = bandwidth calculated as in Section 4.2.2 (Hz)

The interference power is calculated over a given bandwidth in (4.16). From a practical standpoint, the interference spectral power density, $I(f)$, will probably be measured over a set bandwidth by the receiver, anyway. So (4.16) would be reduced to a summation of the interference power over an observed bandwidth, such as during a spectrum sweep period of the cognitive radio.

$$I_{pwr} = 10 \log_{10} \left(\frac{1}{B} \int_{f_c - B/2}^{f_c + B/2} I(f) df \right) \qquad \text{(dBm)} \tag{4.16}$$

The calculation of interference power is different than SINR from an objective perspective: SINR helps the cognitive engine decide if it is good for the waveform to transmit on this frequency. The interference objective looks at the use of the spectrum from the external perspective to see how much overlap exists between competing signals for the same spectrum. Focusing on this objective biases the cognitive engine away from using a waveform that conflicts with another user for the sake of the resources rather than for the capabilities of the waveform.

Looking at interference as a secondary user, instead of using this as an objective function, the cognitive engine could use this concept as a constraint on the objective space. A map of known primary user signals [20] would represent frequencies that are absolutely off limits for transmission. Therefore, the cognitive engine would not simply try to avoid interference-prone spectrum, but it would be forced not to use the spectrum.

4.2.5 Signal to Interference Plus Noise Ratio (SINR)

Dependencies

Knobs: transmit power

Meters: noise power, path loss

Objectives: interference, bandwidth

Definitions

P_T = EIRP (dBm)
L = estimated path loss (dB)
N = noise power as calculated in (4.6) (dBm)
I = interference power as calculated in Section 4.2.4 (dBm)

The signal to interference plus noise ratio (SINR) can inform the cognitive radio about how the presence of interferers can affect the signal reception. Equation (4.6) provides the noise power and (4.16) provides the interference power in the bandwidth of the signal. The received power is then calculated using the estimated path loss from the meters and the transmitter power of the new waveform as in (4.5). The SINR calculation is shown in (4.17) where the noise power and interference power are summed in the linear domain (mW).

$$\begin{aligned} i &= 10^{\frac{I}{10}} \qquad \text{(mW)} \\ n &= 10^{\frac{N}{10}} \qquad \text{(mW)} \\ \text{SINR} &= (P_T - L) - 10\log_{10}(i + n) \qquad \text{(dBm)} \end{aligned} \tag{4.17}$$

4.2.6 Throughput

Dependencies

Knobs: modulation type, symbol rate, number of bits per packet

Meters: none

Objectives: bit error rate

Definitions

l = number of bits per packet (bits)
P_e = bit error rate as defined in Section 4.2.1
R_b = bit rate (bps)
R_s = symbol rate (sps)
k = number of bits per symbol

Throughput is a measure of the amount of good information received. This definition distinguishes throughput from data rate in that data rate is simply a measure of the rate at which data arrives with no consideration for transmission errors. There are many papers available to describe throughput estimations for networks that include retransmission of bad packets. We will circumvent the complexities of that kind of analysis by assuming no retransmissions and furthermore no coding, so a single bit error leads to a packet error. These assumptions are made because the SDR platform used in our experiments does not yet support these capabilities. Although this models the behavior of the given SDR system, the error rate will be much larger than it would be in a system with proper channel coding and protocols. Finally, we assume uniform distribution of bit errors. Once again, it is clear that this formulation does not model all environments and networks properly and could benefit from adding more comprehensive and advanced objective functions.

The probability of a packet error, or the packet error rate, is shown in (4.18).

$$P_p = 1 - (1 - P_e)^l \tag{4.18}$$

For each correctly received packet, l bits are received over a time period of l/R_b where $R_b = kR_s$. On average, then, for a packet error rate of P_p, the bit rate is modified by the probability of receiving a good packet, $1 - P_p$.

$$R_{th} = R_b(1 - P_p) = R_b(1 - P_e)^l \qquad \text{(bps)} \tag{4.19}$$

This objective directly relies on the BER, so using these two objectives together doubly weights minimizing the bit error rate. While this might be a desirable result for some situations, our initial tests on the performance show that the optimization pressure exerted by this objective increases the use of higher-order modulations. This effect helps trade off the solution space between low BER modulations and high data rate modulations. We prefer to keep the objectives as separate as possible in order to more easily adjust selection pressure depending on QoS requirements. In the implementation, we currently use a simplified version of the throughput calculation that only relies on the raw data rate. This objective can be paired with the BER objective in different ratios to provide a proper balance. Equation (4.20) shows the simple throughput objective.

$$R_{th} = kR_s = R_b \qquad \text{(bps)} \tag{4.20}$$

4.2.7 Power

Dependencies

Knobs: Transmit power

Meters: none

Objectives: none

Definitions

P_T = transmit power (dBm)

There are two ways to look at power as a resource. The first way is to think about power in terms of how the radio transmitter uses the external power in the spectrum. In this manner of speaking, power is a shared resource by all radio nodes, so radios should strive to reduce their transmission power. This objective balances efforts to reduce BER or maximize SINR. The transmitted power used here refers to the power of the signal sent to the antenna. In all of the calculations here, though, we have assumed EIRP such as when calculating the received power for the BER equations. In these calculations, we assume a 0-dB gain antenna, so the two are the same. The assumption is based on the lack of the antenna as a knob or even a parameter in the current analysis. In a more developed system that either has a static antenna gain or a smart antenna capable of doing beamforming, the antenna gain would be used here to add to the transmit power as well as in the BER equations to calculate the EIRP.

4.2.8 Computational Complexity

Dependencies

Knobs: modulation type, symbol rate

Meters: none

Objectives: none

Definitions

k = number of bits per symbol
R_0 = symbol rate (sps)

The second way to analyze power is to measure it in terms of power consumption by a radio. Each waveform consumes a certain amount of power that depends on the processes required to transmit and receive information correctly. For example, noncoherent reception requires less processing power than a coherent receiver, which performs the frequency and phase correction, and faster symbol rates require faster processing speed, and therefore more power. The total power consumed includes all aspects of the transmitter and receiver of a waveform, including the transmitter power. However, the last objective in Section 4.2.7 takes care of that in its own way so this second objective does not need to account for it.

As we deal with SDR technology, power consumption maps almost directly to the computational complexity of an algorithm. The mapping between these is not straightforward, especially given the low-power states and techniques available in most processors and hardware. Furthermore, different processors and implementations of the same waveform may consume different amounts of power. Unfortunately, a generalized solution or straightforward mathematical equation to understand or model the power consumption does not exist.

To solve this problem, we are forced to analyze the power consumption in terms of strict computational complexity of the known SDR modes. That is, we break each piece of a waveform down to its components, figure out the computational resources required to run these components, and then put together a lookup table for a given waveform's power requirements. Luckily, the task is simpler than trying to measure all possible waveforms since the great majority of the differences exist within the modulator and demodulator. These discrete values are easily calculated and used in a database for use by the optimizer. The computational analysis for the GNU Radio modulators used here is presented in Appendix C.

The other aspect of computational power is mainly due to the sampling rate required. In this case, the symbol rate directly defines the sampling rate and therefore the computational power required for a waveform.

All other aspects of the waveform involved in this work do not lead to an increase in computational power, though other concepts such as channel coding, source coding, interleaving, or spreading will certainly affect this objective.

4.3 Multiobjective Optimization: A Different Perspective

The previous sections described the objective functions we focus on in this work. When reviewing these, there is overlap as well as interactions among certain objectives. When any one objective is optimized, it affects the performance of some other objectives, either positively or negatively. We find it constructive to graphically represent these interactions to understand the complexity of the multiobjective environment. In Figure 4.2, we list each objective as a node and two types of edges between nodes. Solid edges with arrows define a direct relationship between two objectives, where one objective depends on the other objective being pointed to. The dashed edges with no directivity indicate some indirect dependence through one or more knobs. For example, BER, bandwidth, spectral efficiency, throughput, and computational complexity all depend on the type of modulation used, so changing the modulation affects all of these objectives in some way.

4.4 Multiobjective Analysis

Equation (4.1) provides the basic formula to describe a multiobjective optimization problem, and the preceding sections describe a set of possible objectives when optimizing the PHY layer of a radio. What remains is to understand how the cognitive engine can use these to analyze the behavior of different waveforms and select the best one. The method of performing this analysis is a large factor in the success of a multiobjective optimization problem.

4.4.1 Utility Functions

The most straightforward method of selection is to build a single utility function that combines the objectives into one number. The algorithm can then easily rank the solutions and select the solution that maximizes (or minimizes) the utility function. Utility functions are a core research area in economics and

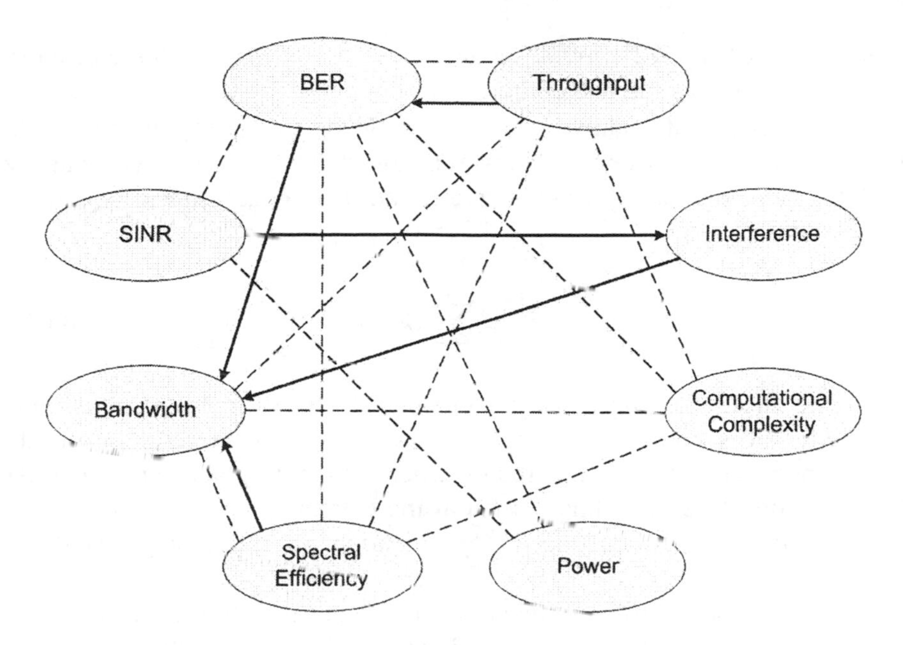

Figure 4.2 Dependency map of objective functions. Solid lines represent direct objective dependency and dashed lines represent indirect dependency through one or more knobs.

operations research. Our intent here is to provide a few possible and popular methods of utility function design and a brief analysis of their properties.

The most basic utility function is the weighted-sum approach, shown in (4.21) and using the definitions from (4.1). In this equation, U is an overall metric of performance. The weights, w_i, applied to each function, f_i, are a weighting of importance, or preference, of the objective.

$$U = \sum_{i=1}^{N} w_i f_i(\bar{x}) \tag{4.21}$$

We have dealt with the concept of the weighted-sum and problems of normalization in [21]. While popular in engineering problems, it has many drawbacks. One of the most significant problems is that this form of utility assumes additivity between each objective, where each objective is independent of the others and can be simply summed together. As discussed throughout and shown in Figure 4.2, the objective functions developed for waveform optimization are not independent. The economic literature on utility functions spends a significant amount of time on this concept and the development of utility functions with respect to the relationship of goods in the analysis. Chung [22] provides a good overview of many of these different functions, which we summarize here.

A development slightly beyond the weighted sum utility function is the linear-logarithmic, or Cobb-Douglas, function in (4.22), where U is the utility, q_i is the value of quantity or objective i, and β_i is an adjustment coefficient for quantity i.

$$\ln U = \sum_{i=1}^{n} \beta_i \ln q_i \tag{4.22}$$

The linear-logarithmic utility function, like the weighted-sum, assumes additivity but is slightly more useful because it shapes the results using the logarithmic function. One important element of the weighted-sum is that the linear relationships often find optima at the extreme edges. The logarithmic function provides convexity to the optimization curve that helps avoid this problem.

In the linear-logarithmic function, β_i is equivalent to the weights w_i in (4.21). The change of notation here is done for consistency with the literature. The weights are important because, as we have discussed, different users and applications have preferences in the quality of service provided by the communications system. This concept leads to the use of consumer preference in economics, where the preferences indicate an indifference curve. The indifference curves represents a trade-off in the amount of any one objective such that the consumer is indifferent to which point is selected. Different users and applications have different requirements, and will therefore be represented by different indifference curves in the objective space. Objectives in the waveform analysis can show both substitutive and complementary affects. For example, an application may be willing to substitute bit errors for higher data rates, but at some point the BER can only go so high before requiring coding, or retransmission, or else the application can no longer handle the errors. At this point, BER becomes complementary to throughput because no data rate can make up for the high numbers of bit errors.

Another well-known function is the constant-elasticity-of-substitution (CES) function of (4.23).

$$U = \left(\sum_{i=1}^{n} \beta_i q_i^{-\rho} \right)^{-1/\rho} \tag{4.23}$$

Here $\rho = (1 - \sigma)/\sigma$ and σ is the elasticity between items, which indicates how much one item can be substituted for another (an elasticity of 1 indicates perfect substitution). This function does not assume the inputs are independent and provides more flexibility to the input values. Because of the dependencies between objectives in the waveform optimization, the CES function suggests a better fit to the representation of utility, which would require an analysis of the elasticity value to use. One potential problem is that this function, as its name says, uses a constant value of the elasticity for all objectives. If one value fits the problem domain, then this is not a problem. However, a more appropriate representation of utility might allow for different values of elasticity between objectives. This suggestion is problematic because then each elasticity value needs to be known and understood, and the value suggests a heavy dependence on domain knowledge, though perhaps this could be learned by the cognitive radio.

Again, we only introduce these functions to show a sample of the functions provided in economic research, but we cannot properly do justice to it here. Much more study is required to relate these two fields and find the most appropriate representation to use in optimization routines of waveforms. Specifically, we need to determine how to develop or derive the coefficients.

4.4.2 Population-Based Analysis

Another method of evaluating performance in a multiobjective problem space is using population-based analysis and Pareto-ranking. We show in Chapter 5 how this concept lends itself nicely to genetic algorithm optimization. Fonseca and Fleming [9] have done a lot of work in this area as well as many other researchers using genetic algorithms for multiobjective analysis.

The Pareto-ranking analysis takes a set, or population, of possible solutions to a multiobjective problem and looks to see which members are nondominated; that is, which members of the population outperform others in all dimensions. We provide a more mathematical definition to this concept in Chapter 5. In Pareto-ranking, each potential solution is ranked relative to other solutions. In a search or optimization algorithm, the idea is to push for better and better solutions until they lie on the optimal Pareto front. This set of solutions represents a trade-off space among all objectives. The final step of the algorithm is to select the solution that best represents the desired trade-off, which is done through some subjective or weighted analysis process to find the proper trade-off space. Park et al. produced recent results on the use of Pareto ranking genetic algorithms for optimization of 3G systems [23].

While multiobjective problem solvers have often used Pareto-ranking, Purshouse and Fleming [24] suggested that this form of analysis works well for a small number of objective functions (two to three) but not as well for larger numbers of objectives, (what they call a "many-objective" problem). Hughes [25] later tested this hypothesis with an experiment to compare different methods of solving multiobjective problems with various numbers of objectives using evolutionary algorithms.

The other two methods for multiobjective analysis suggested by Hughes include multiple single objective Pareto sampling (MSOPS) and repeated single objective (RSO). The MSOPS approach first calculates all of the objectives for each member of the population, and then builds a matrix where the columns represent the objectives, and the rows represent each of the individual's rank for each objective. The individuals' scores are ranked in each objective, giving the highest ranking individual a score of 1, and then counted up over all the members in the population (i.e., the worst performing member is given a rank of the population size). The population members are then assigned a rank according to the number of objectives in which each excels.

The RSO approach evolves a population multiple times using a single, different objective each time. The dominant members from each run are kept for use in the final analysis of the Pareto set.

Hughes' results confirmed his hypothesis that the MSOPS and RSO approaches outperform the Pareto ranking approach. His results also show that the MSOPS approach outperforms the RSO approach. While presenting the evidence, he indicates that this is only proof for the limited problem set he used, and, without mathematical proof, each problem might require a different method. Because each of these methods listed here is applicable to use with a genetic algorithm, the basic formulation developed in the next chapter for the application of a genetic algorithm for waveform adaptation holds. Consequently, it will be interesting in future work to develop and compare different methods of objective analysis.

The remaining issue of performing population analysis is the selection of the best solution from the optimized Pareto front that best represents the desired quality of service. One way to do this would be to utilize one of the utility functions that combines multiple objectives into a single objective to build a ranking scheme, although this concept faces the same challenge of representation and combination of dissimilar but dependent objectives as discussed previously.

Another populations-based method of selection is called *Pareto algebra* introduced by Geilen et al. [26]. Pareto algebra assumes a set of possible nondominated choices that have a preference value related to some application. A good example from their paper is the decision for buying a broadcast program depends on the purpose of the television; whether it is for television viewing, video, or gaming. They then develop an algebra to select the TV that, based on certain properties, best satisfies the family's preference. One of the biggest drawbacks to this technique is the need for a large amount of domain knowledge. The impacts and relationships to the objective space must be known a priori to develop the algebra. This technique could be used in the waveform optimization problem by understanding preference relationships between objectives such as throughput or error rates. However, our intent is to provide a generic platform for the analysis of the interactions and behaviors that have different relationships based on the user preference and situation. Pareto algebra would become more useful to a cognitive engine if it could be generalized and the relationship developed automatically.

There are many concepts of multiobjective ranking and analysis from many different problem domains. We have discussed issues of modeling and analyzing multiple objective problems. We still see a lot to be learned from many of these techniques and benefits from them in the application to the waveform optimization. We use this discussion in the development of the genetic algorithm optimization routine, although the research is ongoing to fully solve this problem. The cognitive engine we discuss will provide the platform by which this problem can be addressed and more fully understood.

4.5 Conclusion

In this chapter, we discussed the concept of multiobjective optimization, both from the theoretical perspectives and regarding the use of this type of analysis to optimize waveforms for a cognitive radio. We only presented a subset of potential objectives we could ask the cognitive radio to meet. New objective analysis can enhance the behavior of the cognitive engine by providing more information from which to make its decisions. Work in this field is already developing useful objective functions. A good example of this kind of work is Marshall [27] who analyzes intermodulation as a metric for deciding free spectrum to use for communicating.

While discussing the possible objectives and defining the mathematical basis for the multiobjective analysis, we point out where improvements in the mathematics may lead to better solutions as they better represent the problem space. This leads to a bigger question of mathematical modeling and uncertainty in cognitive radio systems: how much information is required to make a good decision? Likewise, how much gain comes from more complex analysis? These are interesting research questions beyond the scope of this book, which aim at providing the basic method of analysis and methodology required to perform full waveform optimization in a cognitive radio.

The next chapter deals with the use of genetic algorithms to solve the multiobjective problem. In it, we provide a mathematical definition and analysis of the Pareto-front and Pareto-ranking. We also provide the means by which the current implementation of the genetic algorithm handles waveform optimization.

References

[1] T. Rappaport, *Wireless Communications: Principles and Practices*, Upper Saddle River, NJ: Prentice Hall, 2001.

[2] M. McHenry, E. Livsics, T. Nguyen, and N. Majumdar, "XG Dynamic Spectrum Access Field Test Results," *IEEE Comm. Mag.*, Vol. 45, No. 6, pp. 51 – 57, Jun. 2007.

[3] S. Haykin, "Cognitive Dynamic Systems," *IEEE Proc. Acoustics, Speech and Signal Processing*, Vol. 4, Apr. 2007, pp. IV–1369 – IV–1372.

[4] T. C. Clancy, "Achievable Capacity Under the Interference Temperature Model," *IEEE Proc. INFOCOM*, May 2007, pp. 794 – 802.

[5] J. Neel, R. M. Buehrer, B. H. Reed, and R. P. Gilles, "Game Theoretic Analysis of a Network of Cognitive Radios," *Midwest Symposium on Circuits and Systems*, Vol. 3, 2002, pp. III–409 – III–412.

[6] C. Hwang and A. Syeed, *Multiple Objective Decision Making – Methods and Applications*, New York: Springer-Verlag, 1979.

[7] T. W. Rondeau, B. Le, C. J. Rieser, and C. W. Bostian, "Cognitive Radios with Genetic Algorithms: Intelligent Control of Software Defined Radios," *Software Defined Radio Forum Technical Conference*, 2004, pp. C–3 – C–8.

[8] E. Zitzler and L. Thiele, "Multiobjective Evolutionary Algorithms – A Comparative Case Study and the Strength Pareto Approach," *IEEE Trans. Evolutionary Computation*, Vol. 3, pp. 257 – 271, 1999.

[9] C. M. Fonseca and P. J. Fleming, "Genetic Algorithms for Multiobjective Optimization: Formulation, Discussion, and Generalization," *Proc. Int. Conf. Genetic Algorithms*, 1993, pp. 416 – 423.

[10] J. G. Proakis, *Digital Communications*, 4th ed., New York: McGraw Hill, 2000.

[11] L. W. Couch, *Digital and Analog Communications Systems*, 7 ed., Upper Saddle River, NJ: Prentice Hall, 2007.

[12] T. M. Gallagher, "Characterization and Evaluation of Non-Line-of-Sight Paths for Fixed Broadband Wireless Communications," Ph.D. diss., Virginia Tech, 2004.

[13] R. M. Buehrer, "Equal BER Performance in Linear Successive Interference Cancellation for CDMA Systems," *IEEE Trans. Communications*, Vol. 49, No. 7, pp. 1250 – 1258, Jul. 2001.

[14] F. Jondral, "Automatic Classification of High Frequency Signals," *Signal Processing*, Vol. 9, pp. 177 – 190, 1985.

[15] B. Le, T. W. Rondeau, D. Maldonado, D. Scaperoth, and C. W. Bostian, "Signal Recognition for Cognitive Radios," *Software Defined Radio Forum Technical Conference*, 2006.

[16] D. L. Decker, "Computer Evaluation of the Complementary Error Function," *American Journal of Physics*, Vol. 43, No. 9, pp. 833 – 834, 1975.

[17] M. Abramowitz and I. A. Stegun, *Handbook of Mathematical Functions with Formulas, Graphs, and Mathematical Tables*, New York: Dover, 1972.

[18] S. Elnoubi, S. A. Chahine, and H. Abdallah, "BER Performance of GMSK in Nakagami Fading Channels," *Proc. Radio Science Conference*, Mar. 2004, pp. C13 – 1–8.

[19] K. Murota and K. Hirade, "GMSK Modulation for Digital Mobile Radio Telephony," *IEEE Trans. Communications*, Vol. 29, No. 7, pp. 1044 – 1050, 1981.

[20] J. M. Chapin and W. H. Lehr, "Time-Limited Leases in Radio Systems," *IEEE Comm. Mag.*, Vol. 45, No. 6, pp. 76 – 82, Jun. 2007.

[21] B. Fette, Ed., *Cognitive Radio Technology*, New York: Elsevier, 2006.

[22] J. W. Chung, *Utility and Production Functions*, Cambridge, MA: Blackwell, 1994.

[23] S. K. Park, Y. Shin, and W. C. Lee, "Goal-Pareto Based NSGA for Optimal Reconfiguration of Cognitive Radio Systems," *IEEE Proc. Cognitive Radio Oriented Wireless Networks and Communications*, Aug. 2007.

[24] R. Purshouse and P. Fleming, "Evolutionary Many-Objective Optimisation: An Exploratory Analysis," *IEEE Congress on Evolutionary Computing*, Vol. 3, 2003, pp. 2066 – 2073.

[25] E. J. Hughes, "Evolutionary Many-Objective Optimisation: Many Once or One Many?" *IEEE Congress on Evolutionary Computation*, Vol. 1, Sep. 2005, pp. 222 – 227.

[26] M. Geilen, T. Basten, B. Theelen, and R. Otten, "An Algebra of Pareto Points," *Fundamenta Informaticae*, Vol. 78, No. 1, pp. 35 – 74, 2007.

[27] P. F. Marshall, "Dynamic Spectrum Management of Front End Linearity and Dynamic Range," *IEEE Sym. New Frontiers in Dynamic Spectrum Access Networks (DySPAN)*, 2008.

5

Genetic Algorithms for Radio Optimization

In Chapter 4, we presented waveform optimization as a multiobjective problem. While genetic and evolutionary algorithms are not the only means of solving multiobjective problems, they have continuously proven themselves to excel in this respect [1]. The application of genetic algorithms to multiobjective problems has been the focus of a large part of the research literature, as evident in the proceedings of Genetic and Evolutionary Computation Conference (GECCO) [2]. We use genetic algorithms because of their robust search and optimization capabilities as well as their flexibility in representing the search space and search parameters. Without reproducing the entire body of knowledge on the subject, which we leave to the classic text of Goldberg [3] or the proceedings of GECCO, we start this chapter with a brief introduction to genetic algorithm implementation with a small well-known example. We then present the methods used in the use of genetic algorithms on waveform optimization. Much of this chapter also appears in some of our other published work in [4, 5].

5.1 A Brief Review

In their most simplistic form, genetic algorithms (GA) are single-objective search and optimization algorithms. Common to all GAs is the chromosome definition that determines how the data is represented, the genetic operations of crossover and mutation, the selection mechanism for choosing the chromosomes that will survive from generation to generation, and the evaluation function used to determine the fitness of a chromosome. All of these operations are described in [3].

A genetic algorithm encodes a set of input parameters that represent possible solutions into a chromosome. The evaluation stage calculates a ranking metric of chromosome fitness for each individual, which then determines their survival to the next generation. Optimization progresses through finding genes that provide higher fitness for the chromosome in which it is found. The fitness calculation is often done through some absolute metric such as cost, weight, or value by which the algorithm can rank the success of an individual. Selection is the technique by which more fit individuals are chosen to survive and reproduce for the next generation while less fit chromosomes are killed off. An algorithm terminates when it reaches a desired level of fitness in the population, a single member exceeds a desired fitness, the fitness plateaus for a certain number of generations, or through a simple criteria based on a maximum number of generations. The algorithm then takes the most fit individual of the last generation as the solution.

To understand each of these concepts more clearly, we describe a genetic algorithm that solves the knapsack problem.

5.2 Simple Example: The Knapsack Problem

To explain the operation of a simple GA, we examine the knapsack problem [6], which is a classic NP-complete problem [7] and also called the subset-sum problem (SSP). The knapsack problem takes a set of items, each with a weight and profit, and tries to fit as many of these items into the knapsack to maximize the profit while not exceeding the maximum weight the knapsack can hold. Mathematically, the knapsack problem is represented by (5.1) where the goal is to maximize the profit, P, of items in the knapsack. The problem statement includes K as the maximum weight the knapsack can hold and N_s as the number of items in the set, S. Items in the knapsack have a weight vector, $\bar{w}$, a profit vector, $\bar{p}$, and a the vector $\bar{x}$ is used to show whether an item is present or not in the knapsack. The values w_i and p_i represent the weight and profit of item i, and the knapsack item vector x_i is a 1 if the item is included in the knapsack and a 0 if not.

$$\begin{aligned} P &= \max \sum_{i=1}^{N} x_i p_i \\ \text{subject to: } &\sum_{i=1}^{N} x_i w_i \leq K \end{aligned} \tag{5.1}$$

Chromosome: The problem consists of choosing the right set of items to place in the knapsack, so the chromosome represents the vector x and consists of 1s and 0s, as shown in Figure 5.1, where a 1 indicates that the item is present in the knapsack and 0 indicates that the item is absent.

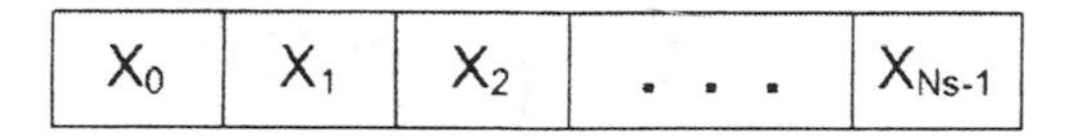

Figure 5.1 Chromosome representation of knapsack item vector.

Selection: Parents are selected based on their fitness in the population, usually with some randomness to allow less fit individuals a chance of survival. The theory is based on Holland's original work on the subject [8] where *schemata* define discrete sections of chromosomes. An overall unfit chromosome may still include one or two highly fit schemata, so preserving some of these and allowing them to create offspring gives those schemata the chance to survive and combine with other good schemata to form a much more fit chromosome in the next generation. Fit parents carrying their genes to the next generation provide *exploitation* of current good genes while allowing unfit members and random schemata into the generational mix provides *exploration* of the search space. These two concepts help define the search capabilities. Good exploitation will help converge quickly to a solution, but the solution may be in a local optimum. Exploration using more random genes helps search wider and farther in the search space, but this can slow convergence. Balancing out how selection takes place as well as properties such as the number of parents to kill during a generation affects the performance of the algorithm but are likewise dependent on the problem space. De Jong [9] analyzed five selection methods in his doctoral dissertation to provide more understanding of this affect.

For the knapsack problem, we use the standard roulette wheel selection technique. Goldberg [3], again, is the source for this, though we will say a few words here to explain the basics. In the roulette wheel selection, the fitness values of all chromosomes are normalized such that the sum of the fitnesses of the population is 1.0. We then imagine putting the fitness values on a roulette wheel in no particular order. The selection method calculates a uniformly random number from 0 to 1. Summing the values clockwise from the top of the wheel, the point where the random number falls corresponds to the chromosome selected. Figure 5.2 illustrates the technique where six individuals have fitness values of 0.15, 0.30, 0.20, 0.25, 0.07, and 0.03. A random value of 0.61 is created as the ranking metric, which lies in the individual with fitness 0.20 as this wheel is shown. The chromosome selected from the roulette wheel is biased towards individuals with large fitness values, but the randomness allows the possibility that any individual can be selected.

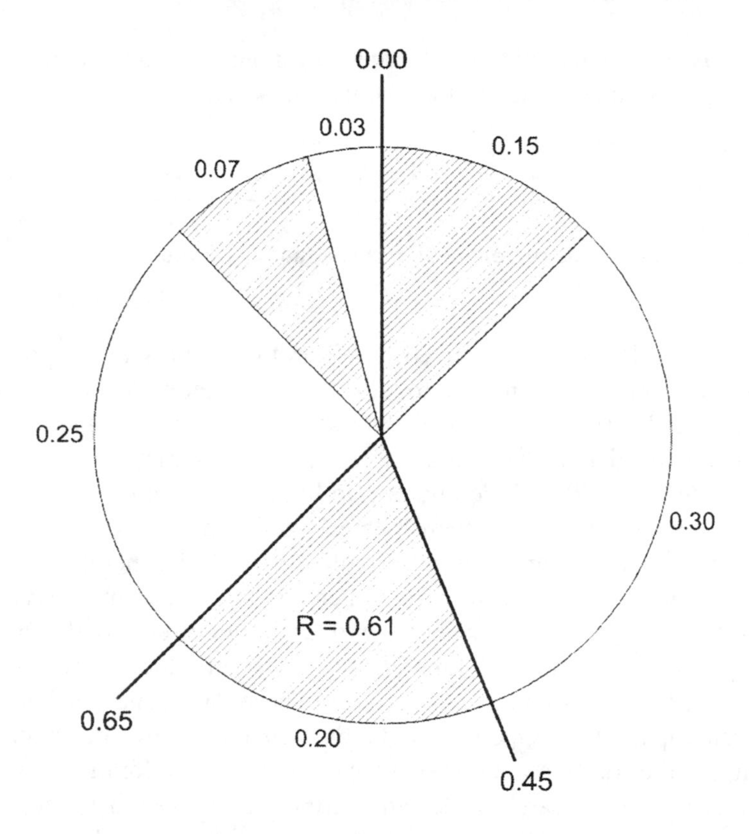

Figure 5.2 Example of roulette wheel selection in a genetic algorithm. Values are counted from the top and added counterclockwise. $R = 0.61$ falls on the item between accumulated values at 0.45 and 0.65.

Crossover: Crossover is performed on two parents to form two new offspring. The GA compares a randomly selected value against a crossover probability that is a property of the GA. If the random number is less than the crossover probability, crossover is performed; otherwise, the offspring are identical to the parents. If crossover occurs, one or more crossover points are generated, which determines the position in the chromosomes where parents exchange genes. Typically, crossover probability is around 0.90 to 0.95, making for high probability of performing crossover. We normally like having a higher crossover probability to keep genes mixing through the population. As we discuss later, we can always choose not to replace all parents with children, which is equivalent to children being exact duplicates of parents. The probability of crossover provides us with another degree of freedom when designing the GA. The number of crossover points used is low and usually

1. Too many crossover points will have too high a probability of splitting the schemata while 1 or 2 points will allow the exchange of genes while preserving long sequences of genes.

Figure 5.3 illustrates the crossover operation of a simple eight-item knapsack problem with two crossover points. The genes after the first crossover point and before the second crossover point are exchanged between parents to form the new offspring.

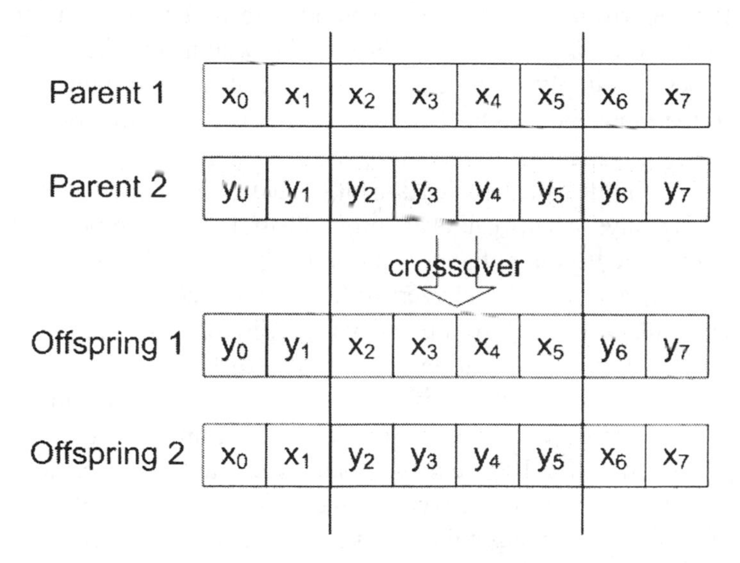

Figure 5.3 Parent chromosomes crossover at points 2 and 6 to create offspring chromosomes.

Mutation: After the offspring are generated from selection and crossover, the offspring chromosomes may be mutated. Like crossover, there is a mutation probability. If a randomly selected real number between 0 and 1 is less than the mutation probability, mutation is performed on the offspring; otherwise, no mutation occurs. Mutation is performed by randomly selecting a gene in the offspring's chromosome and generating a new value based on some probability density function (often a uniform or Gaussian distribution). In the knapsack problem, a gene may be reset to either a 1 or 0 at random. Other techniques simply invert the gene. The probability of mutation is usually set very low, less than 0.10.

Evaluate: Evaluation is probably the most important piece of the GA aside from the initial chromosome definition. Choice of the fitness evaluation is vital to convergence on a highly fit solution. In this knapsack problem, the fitness definition is given by (5.1) where the total profit of the solution is the fitness. The constraint condition tests if the weight of the selected items exceed the maximum knapsack weight. If this occurs, there are a few choices in how the algorithm responds. The algorithm can simply set the chromosome fitness to 0 as a penalty, greatly reducing its chances of reproducing, and thereby removing it from the gene pool. This method has the drawback that the genetic material is completely removed from the population and reduces the overall pool of potential solutions, reduces selection pressure for better genes, and hurts the performance. Another method is to mutate genes in the illegal chromosome until the chromosome meets the weight requirements. In this case, it is best to simply set a gene to 0 instead of random mutation because the goal is to reduce the weight. This keeps genes in the population as potential parents.

Results: Single-objective optimization cases lend themselves to easy performance analysis. The results are often illustrated by a graph like in Figure 5.4 to see how the fitness of the population evolves over the generations. This figure plots the best, worst, and average fitness for each generation.

The results of the GA are shown in Figure 5.4 using a crossover probability of 0.95, a mutation probability of 0.1, and one crossover point. There are 20 members of the population where 15 members are replaced by offspring each generation. The algorithm self-terminated after 5,000 generations. The 100-item knapsack is created by using a uniformly distributed random number between 0 and 1 for both the profit and weight vectors. The maximum weight the knapsack can hold is the sum of the weights of the knapsack items multiplied by another uniform random number between 0 and 1 such that the maximum weight of the knapsack is always less than or equal to the weight of all of the items. Figure 5.4 shows the performance of the best chromosome for three different knapsack problems. Since each knapsack problem has different values for each item and different maximum weights, the curves should be different. The trends, though, are the same with a steep initial rise and asymptotic response towards the optimal value.

Figures 5.5 shows varying results with parameters of the GA with the same knapsack problem. These four tests show differences in performance by changing the population size and the mutation rate. It shows that they all eventually converge to about the same point but at different speeds. Different parameters affect the convergence behavior. In this case, a population size of 20 members with a 20% mutation rate converges fastest, while the larger population and smaller mutation rate converges slowest. This trend suggests that this problem is best solved through random mutations and keeping the population diversity at a minimum.

These figures show a typical trend for GA optimization. The specific values shown in here, however, are specific to the parameters of the GA

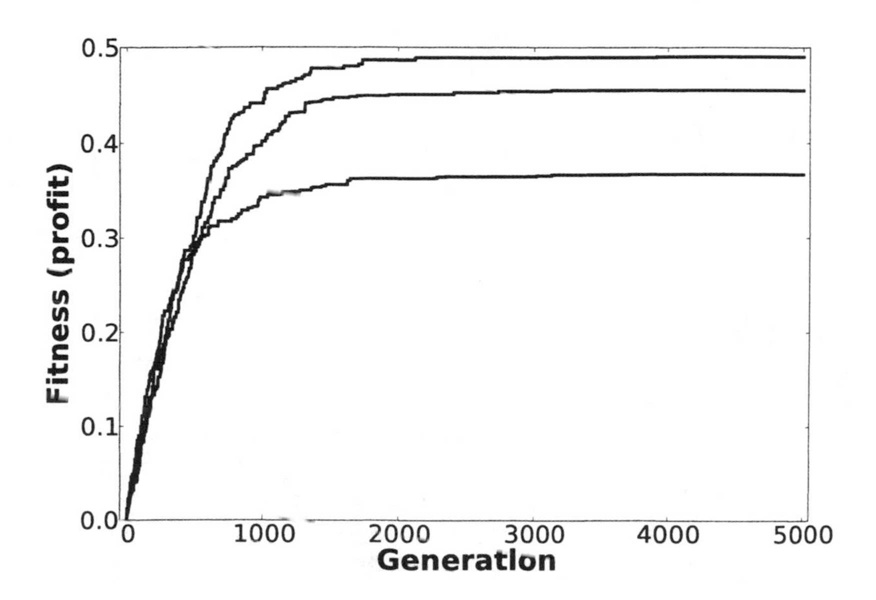

Figure 5.4 Performance graph of three knapsack problems. Each curve represents the best chromosome in each generation for the three different problems.

and the size and values of the knapsack problem. The first generation, which was randomly generated, had poor performance but the best member quickly climbed during the first few dozen generations. A "knee" occurs in the early generations and ends around generation 1,000 where the upwards climb slows down. Over the next couple of thousand generations, the fitness of the best member levels off and does not gain much. Though progress slows down, the fitness still climbs every few generations. At the end, the algorithm is still climbing slowly, and it is likely that the best solution from this generation is not the optimal solution to the problem, although it is likely very close.

In fact, the last point is significant in GA theory. Holland [8] proved that a GA will always converge on the optimal solution; however, it can never guarantee convergence within a certain time limit. Of course, for NP-complete problems, the optimal solution is unknowable through anything other than a full search. Similarly, the optimal solution is not necessarily the goal of real-world optimization problems, which may be interested more in quickly optimized solutions that approach the optimal solution.

The idea of suboptimal (but approaching optimal) solutions is true for the waveform optimization problem where the cognitive engine must produce a waveform under real-time constraints. The cognitive engine needs *better*

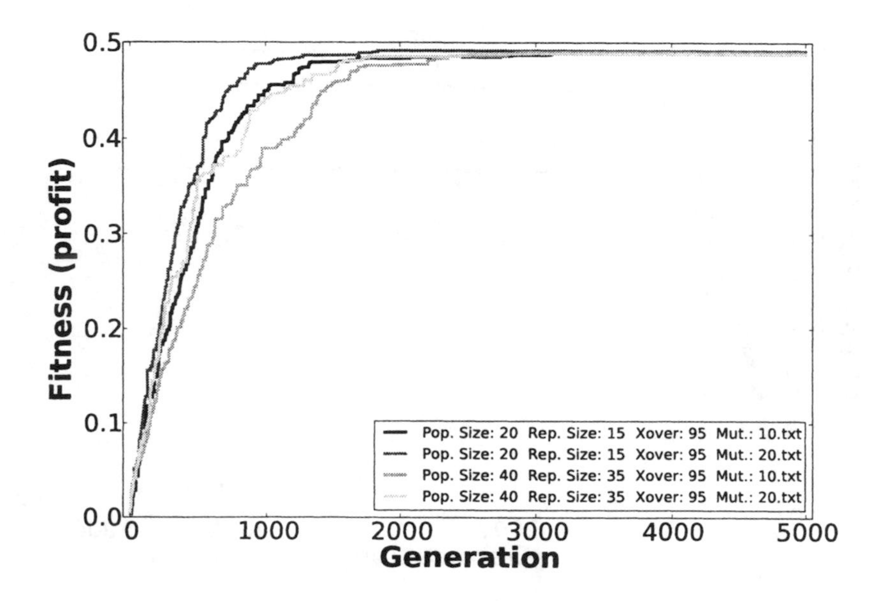

Figure 5.5 Performance graph for the same knapsack problem with different GA parameters.

responses as opposed to a *best* response for the immediate future. Genetic algorithms quickly improve their performance in the first few generations as seen in the performance with the knapsack problem. By generation 1,000, the algorithm has already produced a useful solution. If time and resources permit, the genetic algorithm can search for increasingly better solutions until the answer is required. Techniques like case-based decision theory and distributed computing can further improve performance. We discuss these in later chapters.

5.3 Multiobjective GA

Chapter 4 developed radio optimization as a multiobjective problem and provided a few objectives to use for the optimization. As discussed in Section 4.4, population-based methods are popular for solving multiobjective problems such as the Pareto ranking approach. Population-based analysis lends itself easily to genetic algorithm solvers, and indeed, GAs are well suited to multidimensional optimization due to the parallel evaluation in many dimensions. Genetic algorithms also allow easy implementation of constraints about the problem [10, 11].

To review the argument in Chapter 4, in effective wireless communications, the waveform properties affect the radio's behavior in many dimensions such as BER, bandwidth, power consumption, and throughput rates. Each of these dimensions has some relationship to the QoS, and these relationships change in their relative importance depending on the application being used. For example, large file transfers suggest a need for low BER and high data rate, but a video conference has more demands on delay than BER. Since these goals often compete with each other, as in minimizing BER and minimizing power at the same time, waveform optimization requires joint optimization of many objectives. Genetic algorithms are a powerful approach to autonomously adapting waveforms, a multiobjective genetic algorithm (MOGA).

Different selection and evaluation methods have been proposed for MOGAs [12, 13]. Many methods try to combine the evaluations along the different dimensions into a single metric [14]; this method breaks down in cases where the values of the dimensions can differ greatly in magnitude (BER of 10^{-6} versus data rate of 10^6), and normalizing each dimension requires a large amount of domain knowledge, which might be difficult to obtain [10] or may change over time. Other methods involve competition between population members and incrementing the fitness value of the winner for each objective dominated by the winning member [12]. Horn [15] extends this idea by pairing off two opposing individuals against a larger pool of chromosomes from the population. Each member of the pair competes with each member of the pool. The individual who wins the most competitions against the pool is deemed better fit and survives to the next generation.

We need to introduce a few definitions for the performance analysis before continuing. The general analysis comes down to finding the nondominated solutions in the solution space, known as the Pareto front. These solutions are nondominated when optimization in any objective negatively impacts at least one other objective. That is, a change in a solution cannot simultaneously improve all dimensions at the same time. The Pareto-ranking approach uses the concepts of inferiority and superiority. These definitions assume a minimization problem where $u < v$ means u is more fit than v. We make this clarification because of possible confusion with the terms inferior and superior or when performing a maximization problem. Summarizing from [16]:

Inferiority: $\bar{u}$ is said to be inferior to $\bar{v}$ *iff* $\bar{v}$ is partially less than $\bar{u}$:

$$\forall i = 1, ..., n,\ v_i \leq u_i \ \wedge \ \exists i = 1, ..., n : v_i < u$$

That is, if any of the n objectives of v is less than any objective of u, then u is inferior to v.

Superiority: $\bar{u}$ is superior to $\bar{v}$ *iff* $\bar{v}$ is inferior to $\bar{u}$.

Noninferiority: $\bar{u}$ and $\bar{v}$ are noninferior to one another if $\bar{v}$ is neither inferior nor superior to $\bar{u}$.

Pareto ranking then ranks each member of the population by the number of individuals to which the given member is superior. After some number of generations, the algorithm must return a specific solution, but the population is ranked in terms of Pareto dominance where, ideally, the population represents points along the optimal Pareto front. In this case, if each is Pareto optimal, each population member will be superior to none, and therefore each population will have a ranking of 0. In the end, then, the algorithm needs some method of selecting the member from the final population. We discuss this as we build up the particular implementation of the MOGA for waveform optimization in the next section.

5.4 Wireless System Genetic Algorithm

The wireless system genetic algorithm (WSGA) is a MOGA designed to optimize a waveform by modeling the physical radio system as a biological organism and optimizing its performance through genetic and evolutionary processes. In the WSGA, radio behavior is interpreted as a set of PHY-layer parameters represented as traits or genes of a chromosome. Other general radio functional parameters (such as antenna configuration, voice coding, encryption, equalization, retransmission requests, and spreading technique/code) are also identified as possible chromosome genes for future growth as SDR platforms develop to support each of these traits. Expansion of PHY-layer parameters is a horizontal growth while the MOGA method can also extend vertically to higher layers of the protocol stack such as the MAC or network. Extension to the higher layers will require proper understanding of the objective function analysis of these layers, the genetic representation of the adjustable parameters, and the available communications platform capable of reconfiguration in the layers.

The WSGA makes use of the concepts developed in the last few chapters. The currently available knobs and meters for the simulation environment are given in Table 3.1 and later for the over-the-air experiments in Table 8.22, and the objective functions are given in Section 4.2. The WSGA uses a Pareto ranking selection method similar to [16], but with a few adjustments. First, the WSGA awards points for every objective an individual wins. By doing this, the algorithm has a bit more granularity in how it ranks individuals, especially when two objectives are directly competing, such as BER and power. With these two objectives, the only way to improve or dominate another solution is through a change in the modulation since power and BER are direct trade-offs, so inferiority does not properly allow these objectives to be compared. Second, the WSGA ranks members by the number of members the individual dominates in each objective to make a maximization problem (whereas Fonseca and Fleming [16] rank the individuals by how many members dominate them and thus perform a minimization problem).

Crossover and mutation in the WSGA are simple implementations of these mechanisms. The WSGA uses one crossover point chosen as a uniform random number with a static probability of crossover occurring. The crossover probability is an input parameter to the algorithm as is the number of crossover points, though this is usually kept at 1. Mutation is also a single-point operation chosen from a uniform random number with a static probability of mutation occurring. Future enhancements to the WSGA can include adaptive adjustment of crossover and mutation probabilities as well as the population size during the optimization process for higher convergence efficiency and accuracy.

The use of constraints to a multiobjective problem as shown in (4.1) gives the WSGA the opportunity to incorporate regulatory and physical restrictions during chromosome evolution. If a trait determined by the chromosome exceeds the limits of the radio's capabilities, like finding a center frequency outside the tunable range of the radio, or violates policy by transmitting too much power in a band, then the WSGA applies penalties to the chromosome. A common penalty method for disallowed chromosomes is to set the fitness of the chromosome to zero [17], basically nullifying its chance to survive to the next generation. We found that this reduces selection of good genes that might exist in otherwise disallowed parents. To avoid losing the genetic material, we force random mutations on the gene until it no longer violates the policy, thus preserving large portions of the chromosome structure as well as introducing legal genes into the population. A more focused mutation could also be implemented where mutations occur on the specific genes that cause the chromosome to violate policy, thus not wasting time performing mutations on otherwise acceptable genetic features.

A final implementation problem involves optimization as a network. The WSGA provides a waveform optimized to a single radio node. For the new waveform to be of any use, all other nodes on the network must also use the new waveform. The concept of waveform distribution, issues of optimizing for a number of nodes, and discussion about distributing the processing throughout the network are discussed in Chapter 7.

5.4.1 Details of Chromosome Structure

The WSGA's chromosome structure differs from most traditional GAs because of the variable number of bits used to represent any gene. Most GAs use a single data type or number of bits per gene, but to better represent various radio capabilities, the WSGA uses a flexible representation. For example, a radio might be capable of thousands of center frequencies over multiple gigahertz but only has a few modulations from which to choose. The chromosome can therefore give a large number of bits to the frequency gene and a small number to both the modulation and the transmit power genes, as shown in Figure 5.6. A key result of this structure is that it makes the GA independent of what radio it is optimizing.

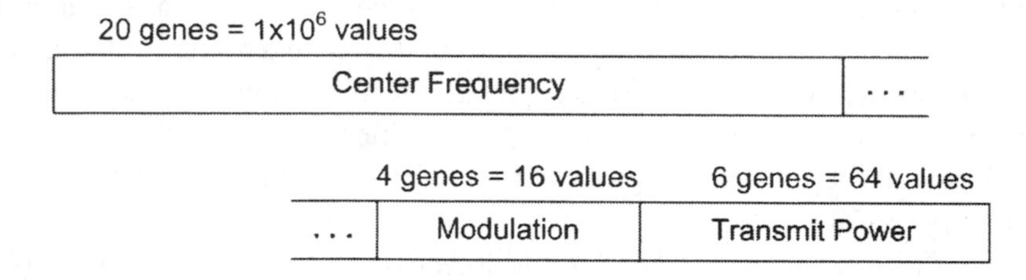

Figure 5.6 Representation of the WSGA's chromosome with variable bit representations of genes.

The variable bit representation is a result of the SDR platform definition file [18]. The platform definition file includes an XML file that looks very similar to the XML listing in Section 2.3.5 to represent the waveform bounds, as well as a DTD file to represent the basic waveform structure. Instead of providing an explicit value for each knob, though, the definition XML file provides the range of values each knob may have. The definition file can be thought of as representing the possible genes in the chromosome while the waveform XML file represents the specific gene. As an example, Figure 5.7 is the representation of the frequency range from a definition file for a radio capable of transmitting from 400 to 500 MHz in steps of 1 kHz and from 2.3 to 2.5 GHz in steps of 100 kHz. It also says that the radio can transmit in the 400 MHz range from 0 to 100 dBm in steps of 0.1 dBm and at 2.3 GHz from 0 to 20 dBm in steps of 0.1 dBm. This XML example shows how both continuous values as well as discrete jumps in values can be represented. The full XML listing is located in Appendix D.

```
...
<Tx>
  <PHY>
    <rf>
      <tx_freq>
        <min unit="kHz">400000< \min>
        <max unit="kHz">500000< \max>
        <step unit="kHz">1< \step>
      < \tx_freq>
      <tx_power>
        <min unit="dBm">0< \min>
        <max unit="dBm">100< \max>
        <step unit="dBm">0.1< \step>
      < \tx_power>
    < \rf>
    <rf>
      <tx_freq>
        <min unit="kHz">2300000< \min>
        <max unit="kHz">2500000< \max>
        <step unit="kHz">100< \step>
      < \tx_freq>
      <tx_power>
        <min unit="dBm">0< \min>
        <max unit="dBm">20< \max>
        <step unit="dBm">0.1< \step>
      < \tx_power>
    < \rf>
...
```

Figure 5.7 Waveform representation in XML.

The XML file provides the bounds and step size, and therefore the number of bits required to represent any possible genetic value for the knob in the chromosome. The DTD file provides the minimum representation of the waveform to structure the chromosome and understand how to parse the XML file. A brief slice of the DTD file is shown in Figure 5.8 while the the full listing is in Appendix D. The WSGA uses the DTD representation to know what genes are available and builds each gene from the XML file above. The logic to accomplish this first parses the XML and DTD files into trees that can be walked. The algorithm removes and elements that contain #PCDATA, which indicates that the element contains data (and is therefore a min, max, or step element). The algorithm then walks the DTD tree looking for leaf nodes on the tree, which are now the elements that describe the tree after the min, max, and step elements are removed. Each of the leaf nodes are names of genes. Each gene has the minimum and maximum value it can contain as well as a step value between the two endpoints. The range represented is easily calculated as $(max - min)/step$ and a $\lceil log_2() \rceil$ of this value provides the

```
<!ELEMENT waveform           (Tx,Rx)>
<!ATTLIST waveform type      #CDATA "analog/digital">
<!ELEMENT Tx                 (PHY,LINK)>
<!ELEMENT PHY                (rf,mod)>
<!ELEMENT rf                 (tx_freq+,tx_power+)>
<!ELEMENT tx_freq            (min,max,step)>
<!ELEMENT min                (#PCDATA)>
<!ELEMENT max                (#PCDATA)>
<!ELEMENT step               (#PCDATA)>
<!ELEMENT tx_power           (min,max,step)>
<!ELEMENT min                (#PCDATA)>
<!ELEMENT max                (#PCDATA)>
<!ELEMENT step               (#PCDATA)>
...
```

Figure 5.8 DTD of waveform representation.

minimum number of bits required to represent the possible values of the gene. Likewise, this information is used in reverse to decrypt the genetic code; the bits representing the gene are an index of $step$ number of steps above the min value. As long as the result is less than max, the gene representation is valid.

When more than one node of the same name exists, an index is used to address the gene. Multiple nodes of this sort are used when continuous ranges do not exist in a radio, such as support for different frequency bands. The number of nodes is indexed into a set of bits and the rest of the gene is made up of the minimum number of bits required to represent the maximum range of any of the nodes. The index position then identifies which range should be used to decrypt the gene.

This method to build chromosomes allows easy representation of a radio's capabilities in XML. The genetic algorithm behaves exactly the same regardless of the radio attached as long as the description is valid in the XML file. The XML and DTD are powerful tools to represent the radio that effectively allow the genetic algorithm to design itself around the radio system, making it truly platform-independent.

5.4.2 Objective Function Definition

XML and DTD files provide the mechanism for generic representation of radio platform capabilities. The next part of the cognitive engine GA implementation is the definition of the objective functions. As discussed in the previous chapter, there are many different objective functions and different implementations of the functions. We presented a few of the objective functions thus far developed as well as a couple of different ways of looking at the functions. The objective functions, like the radio platform, are likely to improve mathematically and representationally over time. It is important, then, that the cognitive engine can be easily adapted to new objective functions.

Again, DTD and XML play their role in this problem along with the use of shared object libraries. The GA is fed the objective functions in the form of an XML file that describes what functions the library holds. The library is compiled as a shared object library that allows dynamic, run-time linking. The GA can then link to the library, pull out the required objectives, and close the library as required. Then, when new objectives are introduced or better mathematical representations are found for the existing objectives, the library can be recompiled separately from the rest of the cognitive engine and uploaded to the cognitive radios for the next optimization process. The XML file that the GA receives contains the list of functions available, so during evaluation, the objective functions are referenced by name in the library. The library processes the objective and returns a solution. The important aspect of this is in the function prototype. Each function representation is in the form:

float $< function\_name >$(radio_hw_def *knobs, radio_meters *meters)

The *radio_hw_def* data structure is a class that contains the information to map the chromosome representation to the radio platform capabilities. The structure is built from the XML and DTD files that define the chromosome as discussed previously. The *radio_meters* is a simple data structure that holds the results of the objective function calculations. Each function returns a real number as part of its result. The WSGA solves a maximization problem, and to allow the generic, dynamic representation of fitness values, this value is a representation of the performance of the objective function as a maximization metric. Simply put, when an objective function should be maximized (e.g., throughput, SINR), the returned fitness is the objective itself. When the objective should be minimized (e.g., BER, power), the returned fitness is the inverse of the objective. These fitness values are then stored as *credits* to represent an individuals fitness. The ranking and selection is based off the values of the credits, which, again, are designed to be compared in a maximization problem. Meanwhile, each member holds a copy of its *meters* data structure used in the post analysis of the algorithm's performance.

5.4.3 Optimal Individual Selection

Pareto ranking only goes so far to provide a population of optimal/near-optimal and nondominated individuals. However, from this point, the algorithm needs to return a single, final individual as the solution to the optimization problem. The Pareto front offers a range of solutions that represent different QoS values. We repeat here an example from [4] with BER and power optimization. Figure 5.9 shows a solution space for bit error rate and power. The optimization goals are to produce a waveform with low BER and low transmit power. Figure 5.9 shows five solutions, labeled A through E, resulting from the availability of three different modulations, BPSK, 8PSK, and QAM16, under SNR conditions of 0 to 12 dB. With the two objectives listed, solutions A, B, and C fall on the Pareto front because any improvement in BER would cause an increase in transmit power. Solutions D and E are, in this case, suboptimal and not on the Pareto front. Any of solutions A, B, or C could be selected based on the optimization criteria, but not all represent certain other properties not specified in the optimization function definition. For instance, BER and power are objectives, yet the quality of service would suffer more from a higher BER than a higher transmit power, so this removes solution A as a candidate. The choice is more narrow between solutions B and C and depends on what property ranks higher; is the extra 1.5 dB power increase worth the decrease in BER from 10^{-5} to 2×10^{-7}? If a third objective such as throughput were added to the optimization problem, solutions D and E are now contenders since they dominate the BPSK modulation scheme in this objective. Again, however, selection from this set is based on quality of service goals by how much importance is given to data rate, power consumption, and BER.

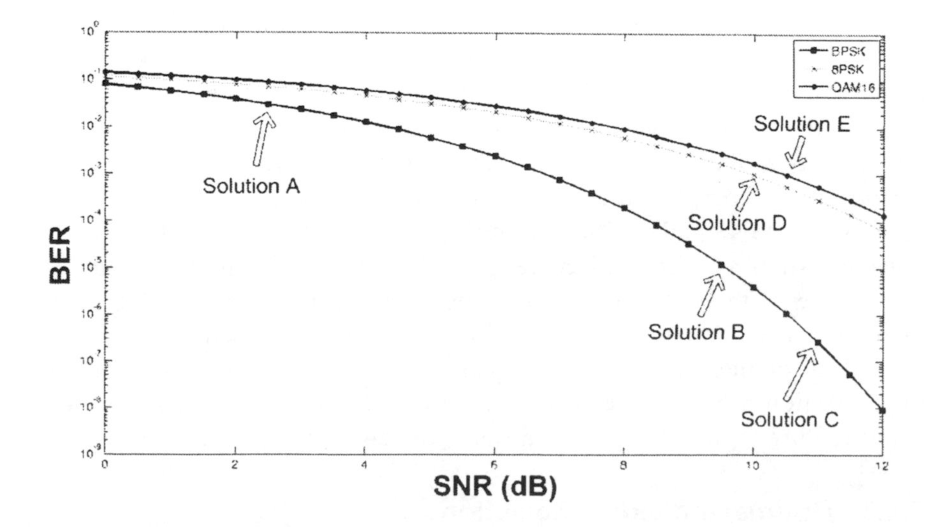

Figure 5.9 Potential solutions for optimization of BER and power; solutions A, B, and C lie on the Pareto front.

As discussed in Section 4.4, there are many different methods for developing an optimization function that provides a ranking method, and therefore the ability in the algorithm to select one individual solution. For the experiments we wish to present, we first normalize the credit values returned from the fitness functions. The normalization is done by keeping track of the maximum value any objective receives throughout the generations. We then use the weights of the objective functions as preference factors in a simple linear-logarithmic utility function of (5.2).

$$f = \sum_{i=1}^{N_O} w_i \ln \left(\frac{c_i}{\lambda_i} \right) \tag{5.2}$$

This equation calculates the fitness of an individual over N_O objectives where each objective has a credit score, c_i, a preference weight, w_i, and a normalization factor, λ_i. The individual in the population with the maximum fitness, f, is selected as the solution of the algorithm for implementation on the radio.

This sets up the design of the cognitive engine's WSGA. Chapter 8 provides examples of of this in operation, after the rest of the cognitive engine design is discussed.

5.5 Conclusions

With the multiobjective waveform optimization problem as well as the range of possible knobs and meters available on a radio platform, the genetic algorithm is a solution to both effective optimization and generic representation to make it an excellent choice to use for cognitive radio work. In this chapter, we presented the basics of what makes the GA such a powerful search and optimization tool as well as explained how it was developed for the cognitive radio problems. The particular aspects of the GA implementation include the generic representation and definition of the chromosomes to allow it to operate on different systems, the equally flexible definition and implementation of the objective functions, and the methodology behind performing the GA operations and making selections. As we discussed, advances remain to be made in the evaluation functions as the relationships are better understood and developed. In Chapter 8, we show both the successful implementation of the WSGA as well as discuss the symptoms of the performance that some of the advancements could correct. Before presenting the operational cognitive engine or WSGA, the next couple of chapters finish the necessary theory and design work required for the full cognitive engine.

References

[1] E. J. Hughes, "Evolutionary Many-Objective Optimisation: Many Once or One Many?" *IEEE Congress on Evolutionary Computation*, Vol. 1, Sep. 2005, pp. 222 – 227.

[2] M. Keijzer, M. Cattolico, D. Arnold, V. Babovic, C. Blum, P. Bosman, M. V. Butz, C. Coello, D. Dasgupta, S. G. Ficici, J. Foster, A. Hernandez-Aguirre, G. Hornby, H. Lipson, P. McMinn, J. Moore, G. Raidl, F. Rothlauf, C. Ryan, and D. Thierens, Eds., *Proceedings of the 8th Annual Conference on Genetic and Evolutionary Computation*, Seattle, Washington: ACM SIGEVO, Jul. 2006.

[3] D. E. Goldberg, *Genetic Algorithms in Search, Optimization, and Machine Learning*, Reading, MA: Addison-Wesley, 1989.

[4] B. Fette, Ed., *Cognitive Radio Technology*, New York: Elsevier, 2006.

[5] T. W. Rondeau, B. Le, C. J. Rieser, and C. W. Bostian, "Cognitive Radios with Genetic Algorithms: Intelligent Control of Software Defined Radios," *Software Defined Radio Forum Technical Conference*, 2004, pp. C–3 – C–8.

[6] R. Spillman, "Solving Large Knapsack Problems with a Genetic Algorithm," *IEEE Proc. Systems, Man and Cybernetics*, 1995, pp. 632 – 637.

[7] M. R. Garey and D. S. Johnson, *Computers and Intractability: A Guide to the Theory of NP-Completeness*, New York: W. H. Freeman & Company, 1979.

[8] J. Holland, *Adaptation in Natural and Artificial Systems*, Boston: MIT Press, 1975.

[9] K. A. De Jong, "An Analysis of the Behavior of a Class of Genetic Adaptive Systems," Ph.D. diss., Univ. of Michigan, 1975.

[10] E. Zitzler and L. Thiele, "Multiobjective Evolutionary Algorithms - A Comparative Case Study and the Strength Pareto Approach," *IEEE Trans. Evolutionary Computation*, Vol. 3, pp. 257 – 271, 1999.

[11] T. Hiroyasu, M. Miki, and S. Watanabe, "Distributed Genetic Algorithms with a New Sharing Approach in Multiobjective Optimization Problems," *IEEE Proc. Congress on Evolutionary Computation*, Vol. 1, Jul. 1999, pp. 69 – 76.

[12] J. D. Schaffer, "Multiple Objective Optimization with Vector Evaluated Genetic Algorithms," *Proc. Int. Conf. Genetic Algorithms*, 1985, pp. 93 – 100.

[13] P. Fleming, "Designing Control Systems with Multiple Objectives," *IEE Colloquium Advances in Control Technology*, 1999, pp. 4/1 – 4/4.

[14] C. Hwang and A. Syeed, *Multiple Objective Decision Making - Methods and Applications*, New York: Springer-Verlag, 1979.

[15] J. Horn, N. Nafpliotis, and D. E. Goldberg, "A Niched Pareto Genetic Algorithm for Multiobjective Optimization," *IEEE Proc. Conf. on Evolutionary Computation, IEEE World Congress on Computational Intelligence*, Vol. 1, 1994, pp. 82 – 87.

[16] C. M. Fonseca and P. J. Fleming, "Genetic Algorithms for Multiobjective Optimization: Formulation, Discussion, and Generalization," *Proc. Int. Conf. Genetic Algorithms*, 1993, pp. 416 – 423.

[17] ——, "Multiobjective Optimization and Multiple Constraint Handling with Evolutionary Algorithms - Part I: A Unified Formulation," *IEEE Trans. Systems, Man, and Cybernetics*, Vol. 28, pp. 26 – 37, 1998.

[18] D. Scaperoth, B. Le, T. W. Rondeau, D. Maldonado, C. W. Bostian, and S. Harrison, "Cognitive Radio Platform Development for Interoperability," *MILCOM*, Oct. 2006, pp. 1 – 6.

6

Decision Making with Case-Based Learning

Decision making is a complex part of the cognitive radio design. A cognitive radio uses environmental and behavioral information about radio performance or user requirements to make decisions on how to adapt. Decisions can include what parameters to adapt, if adaptation is required, or even the method by which to adapt. Our goal in this chapter is to introduce one particular use of decision making in cognitive radios: augmenting optimization through past knowledge. Given changes in the environment, the decision-making system uses past knowledge to aid the genetic algorithm optimization process by providing goals and by seeding the population to best reach the desired goals. The basics of this concept were first published in [1].

Goldberg introduces the concept of knowledge-based techniques in his book [2]. Here, he cites the use of knowledge in how people solve problems: humans do not develop everything from first principles over and over again. Instead, previous knowledge influences and augments current decisions. The discussion tends towards the use of previous solutions to enhance the next routine, but this is not Goldberg's use of the the technique. Instead, he discusses how understanding the problem domain can lead to tailoring the algorithm to help it solve the problem. His techniques involve the use of crossover operations that are specifically tailored to a particular problem. He gives an example of using a crossover operation that preserves legal solutions when solving the traveling salesman problem. Another powerful approach is to use a hybrid system of a genetic algorithm along with another algorithm that does local optimization. In this approach, the genetic algorithm (GA) converges on an area where a solution is likely to exist. Instead of spending generations to lock into the highest point in the search area, a local optimization routine takes over to finish the job. This concept works because local optimizers generally have well-understood, tractable performance, and they lock on optimum points in a local search space quickly. The GA performs the global search and the local optimizer finishes it off.

Ramsey and Grefenstette provided an initial analysis of case-based learning for genetic algorithm population initialization [3]. Their aim was to develop a system that would enable what they call *anytime learning* in changing problem spaces. This goal is similar to the online learning of the cognitive engine. Recently, Newman [4] showed how using previous solutions can improve the performance of waveform optimization. In this chapter, we add to this research through the discussion and implementation of case-based decision theory to illustrate the use of knowledge in solving problems with genetic algorithms. We provide an example of this technique and discuss many advances case-based systems offer. The use of the technique in the cognitive engine is part of the experiments of Chapter 8.

6.1 Case-Based Decision Theory

Our use of decision-making theory is largely derived from the case-based decision theory (CBDT) work of Gilboa and Schmeidler [5]. CBDT uses past knowledge to make decisions about future actions. Case-based decision theory is closely related to CBR [6], and to avoid arguing semantics between the two techniques, we will generically refer to these techniques as case-based learning.

Formally, case-based learning defines a set of problems $q \in P$, a set of actions $a \in A$, and a set of results $r \in R$. A case, c, is a tuple of a problem, an action, and a result such that $c \in C$ where $C = P \times A \times R$. Furthermore, memory, M, is formally defined as a set of cases c currently known such that $M \subseteq C$.

The cognitive radio uses sensors to observe when the environment or user's needs change. This new information is modeled as a new problem, p to be solved by the cognitive radio. The sensors could indicate a change in the interference environment, a new propagation channel, or a change in the application of the radio requiring different QoS needs. The cognitive engine must then determine the action, a, with which to respond. The case-based system analyzes the new problem against past cases in memory to determine the similarities between the new problem and past problems as well as the utility of the past actions. Utility refers to how successful an action was at responding to the problem. The action defined by the current cognitive engine is the waveform to use in the current situation. As the cognitive engine processes and learns, it populates the knowledge base with more cases that better reflect the environment to help make better choices. This technique is similar to an expert system, but one that learns autonomously.

A similarity function defines how similar two cases are and is represented by (6.1). The similarity function is any function that provides some measure of how close two problems are to each other where 0 represents no similarity and 1 represents a perfect match.

$$s : P \times P \rightarrow [0, 1] \tag{6.1}$$

The utility analysis of the past cases is represented in (6.2), which is any function that produces some real-valued result measuring the utility of the action.

$$u : R \rightarrow \Re \tag{6.2}$$

Case analysis comes down to which case is both most similar to the new problem as well as how successful the action was in the past. The decision maker then uses a final decision function to decide which case to use. The simplest implementation is a similarity-weighted decision function as shown in (6.3). A particular problem may be very similar to previous problems in the case base, but the solution to the previous problem might have performed poorly in the past. In this situation, a less similar but better performing case is selected instead.

$$U(a) = s(p, q)u(r) \quad \text{where} \quad (q, a, r) \in M \tag{6.3}$$

This equation is only one function used to make the decision. The challenge of this technique is to create effective similarity, utility, and decision functions that best represent the types of information received through the sensors. We develop this concept further throughout this chapter.

6.2 Cognitive Engine Architecture with CBDT

The cognitive engine uses case-based decision theory to augment the optimization process. Instead of relying on pure optimization alone, the case base helps prime and direct the optimization with learned experience. Likewise, instead of basing all decisions on past actions from the case base, the optimization process allows online learning to build knowledge. The case base and optimization routines work together to enable learning and adaptation in the cognitive engine.

The case base holds past cases, actions, and results of the actions. In the cognitive radio, the case represents some model of the environment, such as a sequence of meter readings or an interference map. The action for a given case is in the form of the waveform created to meet the case's needs. The results, then, are a measurement of how well the action performed.

To develop the performance measurements, the cognitive engine uses the predicted results of the optimization process and analyzes how closely those results match to the actual performance of the radio. The optimization process develops the waveform based on a set of mathematical models in the form of objective functions. The results of the objective functions are calculated performance measures of the waveform. When the waveform is then used in the environment, the resulting performance may differ from the calculated performance. This difference relates to the utility of the waveform.

Figure 6.1 presents the block diagram of the described system. An incoming problem is matched against the cases through a similarity function while the case results are compared to the radio performance to develop the utility of the case. The decision function is an equation like (6.3) that uses similarity and utility to properly select the case most representative of the new problem. The results are then passed to the optimization process along with the new problem. Both the waveform solution and the objective functions' results are fed back along with the problem model to the case base to be stored as a new case.

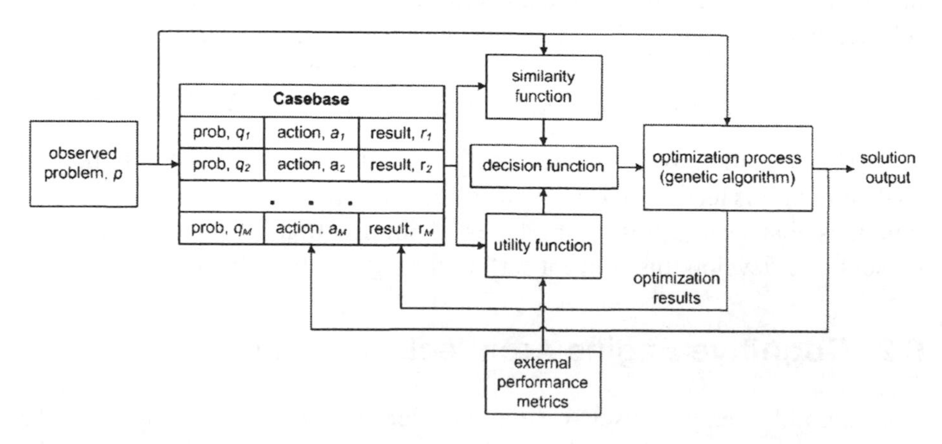

Figure 6.1 Case-based decision theory implementation with optimization process.

This figure and the discussion about the use of case-based decision theory do not focus on any particular optimization process and should be considered a generic method for learning and optimization. In our cognitive engine work, the particular form of the optimization process is a genetic algorithm.

6.2.1 Memory and Forgetfulness

The case base holds M cases, and therefore must have a system to delete, or forget, cases no longer used. Given a full case base, when the cognitive engine observes a new case, either the new case is not remembered or it must replace a current case. We list here a few forgetfulness functions.

Temporal forgetfulness: In this method, the oldest case is forgotten. This method is very simple to implement as a first in first out (FIFO) buffer. Each new case is pushed onto the front of the queue and the oldest case is popped off.

Maximum distance forgetfulness: If the similarity function defines a distance between two cases, a linear relationship can determine which case to forget. Here, a one-dimensional distance determines how similar cases are where distance is measured as $d(x_i, x_j)$. Take the constructed example in Figure 6.2 where the xs represent known cases and the o represents the new case. First, the most similar case to the new case o is found to be x_3. Next, the two cases surrounding these two cases is found, x_2 and x_4. The goal is to maximize the distance between the surrounding cases to provide better coverage of the search space. The case that satisfies (6.4) provides the maximum distance between both cases x_2 and x_4. Edge cases are easy as they are just a maximization of the similarity space.

$$\hat{c} = \min_{c \in [o, x_3]} \{abs\left(d\left(x_2, c\right) - d\left(c, x_4\right)\right)\} \tag{6.4}$$

The new case increases the distance between itself, x_2, and x_4 over the old case x_3, so x_3 is forgotten. The theory behind this approach is to maximize the possible problem space represented by the case base. After replacing x_3 with o, more of the problem space is covered.

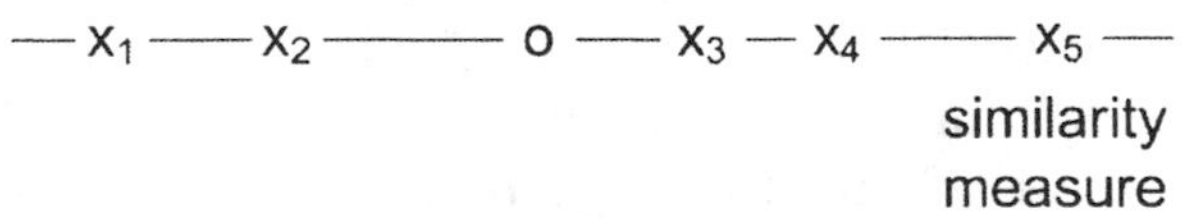

Figure 6.2 Maximizing distance between cases.

Maximum utility forgetfulness: This technique replaces the case with the lowest utility with the new case.

Each of these techniques make assumptions on the problem space such as temporal properties and radio behavior. If the environment changes quickly, then forgetting the oldest case might work well as the case base tracks the changes. If the environment does not change quickly, or if there are certain environmental models that are characteristic of many problems, it is useful to keep many of these models around and not drop one simply because it is old. The ability to properly model similarities and understand utility is also important to the case base system; perhaps there is an environment where no waveform will behave successfully, or it is difficult to build one that does. In this situation, forgetting a case based on low utility might not allow the system enough time to learn the proper response by starting from a blank slate each time.

It is also possible to mix these systems where information is stored in different case bases for different purposes. Rieser discusses this concept in his dissertation [7]. He uses the concepts of short-term memory and long-term memory, where each represents a different method of remembering and forgetting information to take advantage of the different properties each has.

6.3 Cognitive Engine Case-Based Decision Theory Implementation

Figure 6.1 provides the system diagram for how the case base is used with the optimization process. Looking back at Figure 2.3, the new case is received by the cognitive controller through a sensor. The cognitive controller calls the decision-making process to find information in the knowledge, or case, base and then sends the information to the optimization routine for processing. Each of these components can be developed and implemented independently. We introduce here the concepts behind the components for the cognitive radio. We then build on this foundation to solve knapsack problems using CBDT. The ability to move between the knapsack and the cognitive radio illustrates the problem-independent nature of the cognitive engine.

The sensors and genetic algorithm optimization components in the cognitive engine have already been discussed. Chapter 2 discussed the format of the information from the sensors, and Chapters 3, 4, and 5 discussed how the GA optimizes waveforms. The design of the case base is covered here.

The case base is implemented as a relational database in MySQL. Structured Query Language (SQL) is both well-known and well-supported, and the MySQL implementation has proven performance, stability, and design. Integration of a MySQL databases is possible in almost any contemporary programming language, and it reduces the design time of building a new database or case base structure. Another key aspect of the MySQL database is that these database servers are easily accessed over a distributed network, making it easy to share knowledge between cognitive radios.

The case base is structured as a single database with multiple tables. Each sensor is assigned a main table where each row represents a case. The columns of this table are *timestamp*, *problem*, *solution*, and *result*. The *problem* column references another table that describes the problem. Similarly, the *result* column references a table to describe the result's properties. The final piece of the case definition is in the *solution*, which is a text field in the main table.

The references to the problem and result tables are done in order to maximize the flexibility when defining the information representation. Each sensor contains data that will require special representation in the database to maximize the ability of the case base to perform the similarity functions. Likewise, the results from the optimization process may change, and, again, the flexibility in the representation is important for the system to adequately grow. Figure 6.3 shows the database structure.

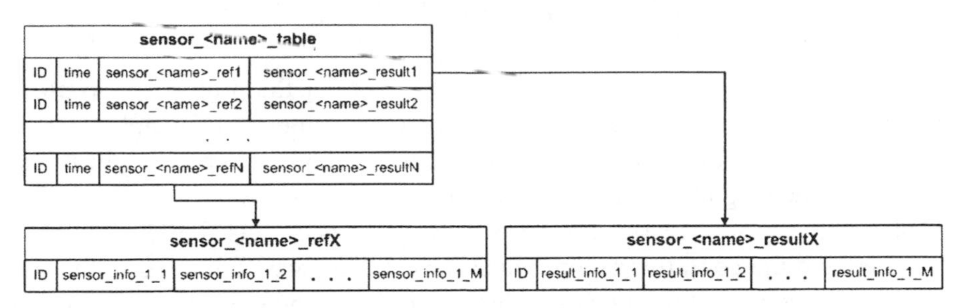

Figure 6.3 SQL database design for the cognitive engine.

The structure of each of the tables is conveyed through a DTD file. As all the information is passed via XML formatting, the DTD represents how the information is presented and therefore stored in the problem and results tables. When the cognitive controller associates with a sensor or the optimization routine, the component passes a DTD representation of the data format, which is used to build the table structure. The XML and DTD formats and uses are described more thoroughly in Chapter 8 and the file formats are shown in Appendix D.

In the current implementation, the cognitive engine associates with three sensors to collect the meters, PSD, and objectives. The wireless system genetic algorithm does the optimization. The PSD sensor provides an interference map that the GA uses in analyzing its objectives, and the case-based decision maker uses the meters sensor in its utility calculation. The objectives are passed through a sensor to define which objectives to use and their associated weights that define the problem space. The results stored in the case base are retrieved from the genetic algorithm optimizer as shown in Figure 6.1. The results are the theoretical, or mathematically calculated, values of the objectives such as bit error rate, SINR, and throughput. The meters sensor calculates the objectives of the waveform's performance.

The utility is a representation of the difference between the calculated objectives and the actual objectives as shown in (6.5). We use the CES utility function because of the easy flexibility it offers to defining the relationships between the quantities used in the utility calculation. Here, there are N_O defined objectives, $f_i(q)$ is the calculated value of objective i for case q, and $f_i(m)$ is the observed value of objective i as derived from the meters. The absolute value is used in this calculation because the waveform is penalized for both better and worse performance than the estimated values. In other words, penalties occur for both over- and underoptimization.

$$u(q) = \left(\sum_{i=1}^{N_O} (|f_i(q) - f_i(m)|)^{-\rho} \right)^{-1/\rho} \qquad q \in P \tag{6.5}$$

The similarity function determines how similar the problem is to other cases in the case base. Problems are defined using the set of weights of the objective functions that determine what level of QoS a user or applications requires. The cognitive engine is designed to produce a waveform that improves the quality of service, so the problem space sets the level of QoS required and the cognitive engine must develop the waveform. Selecting the case comes down to finding a case that is similar to the QoS problems posed in the past that have performed well. The similarity function is then the sum of the differences between the objective weights of the new problem and the weights of the objectives in each case.

$$s(q,p) = \begin{cases} 0, & \text{if } w_i(p) = w_i(q) = 0 \\ w_i(q), & \text{if } w_i(p) \neq 0,\, w_i(q) = 0 \\ 1 - \sum_{i=1}^{N_w} \frac{|w_i(q) - w_i(p)|}{w_i(p)}, & \text{else} \end{cases} \tag{6.6}$$

$$for \quad q \in P$$

In this equation, there are N_w weighted objectives, and each objective i in case q has a weight $w_i(q)$, and the new problem statement p has a weight for each objective $w_i(p)$.

To extend this concept, each sensor could be queried in this manner to find cases from each domain that represents the problem. In the case of the energy detector sensor, it might be useful to determine how similar this interference environment is to past environments to help find white spaces quicker. A potential similarity function for this sensor is a normalized cross correlation between the old interference maps and the new map.

6.4 Simple CBDT Example

To demonstrate the use of case-based decision theory with a genetic algorithm, we revisit the knapsack problem used as an illustration in Chapter 5. Before moving into the use of CBDT on a knapsack problem, Figure 6.4 shows the average fitness per generation of 100 runs of the GA on the same knapsack problem. The knapsack problem and GA parameter settings are identical to those in the discussion in Chapter 5. We average the results of the best chromosome per generation over 100 runs to provide a characteristic performance curve for the genetic algorithm.

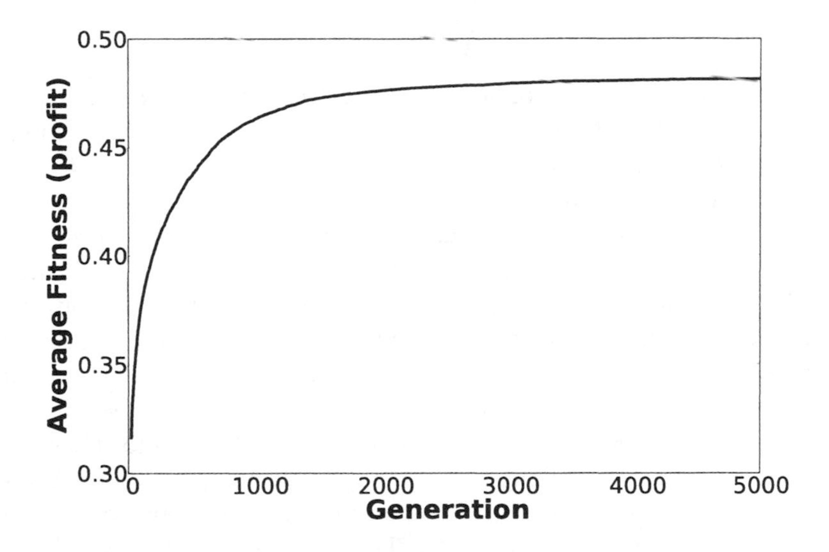

Figure 6.4 Characteristic performance of the knapsack GA averaged over 100 runs.

To apply CBDT to the knapsack problem, the system requires the similarity, utility, and decision functions. The similarity function is defined in (6.7). In this equation, the knapsack problems are modeled by a weight-adjusted profit value, and before the calculation is performed, the vectors of weight-adjusted profits of the N_s items are sorted in ascending order to make a fairer comparison. In the following equations, recall that the cases stored in the case base are identified as case q while the observed problem is p. Each case has a profit and weight vector where the profit of item i is $p_{q.i}$ and the weight of item i is $w_{q,i}$.

$$s(p,q) = 1 - \sum_{i=1}^{N_s} \left| \frac{p_{p,i}}{w_{p,i}} - \frac{p_{q,i}}{w_{q,i}} \right|, \quad q \in P \tag{6.7}$$

We tested two utility functions. Utility function 1 in (6.8) is just the profit of the solution, called the *profit-only* utility function. Utility function 2 in (6.9) makes the assumption that the better solutions will also be closer to filling the knapsack completely, and so the profit is adjusted by the ratio of the weight of the items in the knapsack to the maximum weight the knapsack can hold. This equation is called the *weight-profit* utility function.

$$u(q) = p_q^m, \quad q \in P \tag{6.8}$$

$$u(q) = \frac{w_q^m}{W_q} p_q^m, \quad q \in P \tag{6.9}$$

In this equation, p_q^m is the total profit represented by a solution in the case base, w_q^m is the total weight of the solution, and W_q is the maximum weight of the knapsack for the problem in case q.

The decision equation is the simple form from (6.3).

The first test is for proof-of-concept, debugging, and stability tests of the system. In this experiment, we use the same knapsack model and repeatedly run the GA using individuals initialized by the case base where the case base is initially empty. The experiment runs the GA for 100 generations each time and stores the best performing individual in the case base. The maximum number of cases retrieved from the case base is 10, and the case base can hold a maximum of 100 cases when the oldest case is dropped for the newest case. In the first run, with nothing in the case base, no population initialization takes place. The next run, using the same knapsack problem, looks in the case base and selects the only model there, initializes one individual in the population, and again, stores the best performing individual after 100 generations. Each time through, more members are retrieved from the case base until 10 individuals are stored. At this point, the member with the highest similarity and utility is selected using the above equations along with its nine closest neighbors. The system is run 50 times to produce the same number of generations used to produce Figure 6.4. We averaged the results after 10 trials. Figure 6.5 compares the performance of the CBDT-GA with the normal GA over the same number of generations.

The idea here is to make sure the system performs the analysis and population initialization properly, so we are not expecting to see an overall

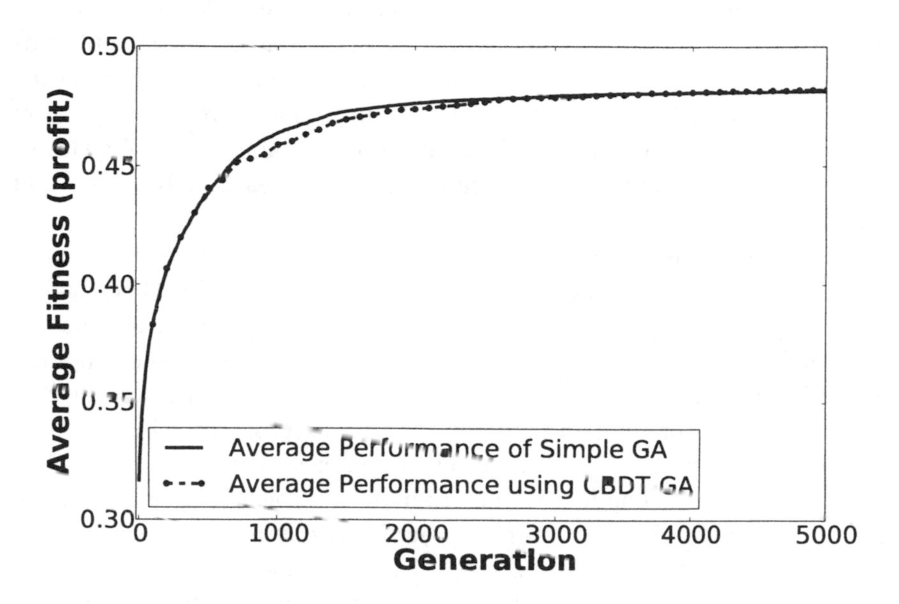

Figure 6.5 Case-based initialized GA compared to the characteristic performance of the standard GA.

performance improvement from this. Because the experiment used the same knapsack problem each time, the similarity calculation is always 1. Each time through the GA, the best performing solution is inserted into the case base; therefore, the utility increases with every case. The case selected from the case base is then always the last case run along with previous nine cases. Effectively, this experiment is the same as running one GA over 5,000 generations and inserting a few random individuals every 100 generations. This is a form of *migration* discussed in the GA literature. For smaller generations, the migration of random individuals into the population affects the performance by removing other members that had been evolving towards the global optimum. However, in later generations, when exploitation of the parents is no longer the driving force of the optimization, the random members help add a search capability to the population to improve its performance. The results in Figure 6.5 confirm the theoretical performance by showing a slight dip in fitness between generations 500 and 2,500, and a slight improvement overall in the performance during the final generations.

The real experiments to see the performance capabilities of the knapsack GA come from analyzing the GA against a number of random knapsack problems. CBDT offers a number of variables to test, so we provide here an experiment to analyze a few of the more relevant issues. We run with the two utility functions of (6.8) and (6.9). Another variable is the number of individuals initialized from the database, so the experiment here looks at the performance of initializing 0 individuals to the maximum number of members of the population (20 in this case). The final variable under test is the size of the case base, where we run the experiment for a case base of 100 cases and 500 cases.

The experiments all used the same GA described previously, and each run consists of 100 generations. For each change in the CBDT variables, we use the same knapsack models. One hundred knapsack problems are first created and stored in files. The case-based system is an implementation of the cognitive engine with a sensor attached that can either create new knapsack problems or select a predefined knapsack problem. The cognitive engine is also associated with the optimization process that runs the knapsack genetic algorithm. A simple Python program controls the sensor and the cognitive controller while the optimization GA is run as a separate process. The Python program initializes the knapsack sensor and cognitive controller, calls M random knapsacks from the sensor, and then runs the optimization process for each knapsack model, where M is the size of the case base. The random knapsack results are inserted into the case base. This randomized set is used to compare the performance of the CBDT-GA in a random environment and to not bias the solutions toward known models. After the case base is randomized, performance statistics are collected by running the simulation with the 100 predefined knapsack models. This collection is done for population initialization of 0 individuals to the population size (20).

Because each knapsack is different, it is important to compare performance between the same knapsacks. Some knapsack problems are harder than others, especially those with smaller maximum weights. The analysis is then to look at the percent improvement over the base case defined for when no individuals were initialized. We average the improvements over all 100 knapsacks to see how well the performance was on average. The following figures show the percent improvement and average performance of each of these experiments.

Figure 6.6 compares the average percent improvement over not using initialization for the four test cases. On average, when M is 100 using the profit-only utility function of (6.8), the improvement is 5.50% and when using the weight-profit utility function of (6.9), the improvement is 4.67%. When M is 500 using the profit-only utility function, the improvement is 4.34% and when using the weight-profit utility function the improvement is 5.94%.

Figure 6.7 shows the performance while using (6.8) with a case base of 100 items. Figure 6.9 shows the performance while using (6.8) with a case

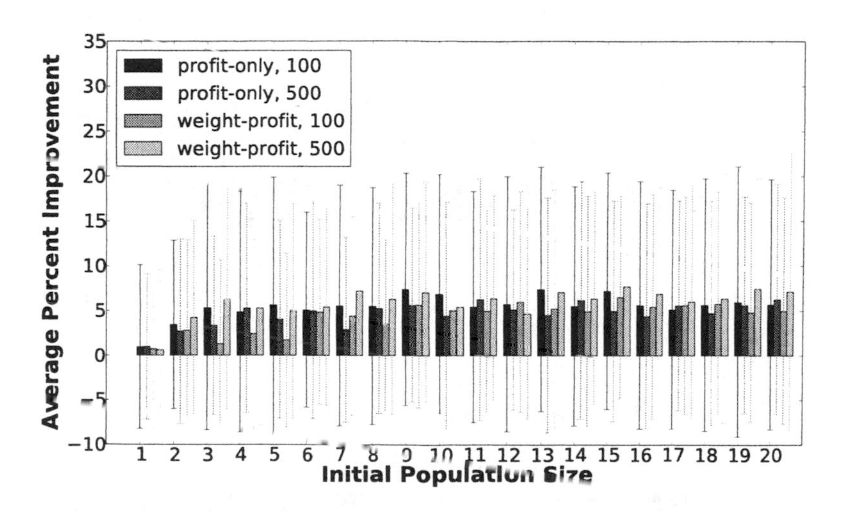

Figure 6.6 Average percent difference in performance of CBDT-GA to no initialization (error bars indicate 1 standard deviation from the sample mean).

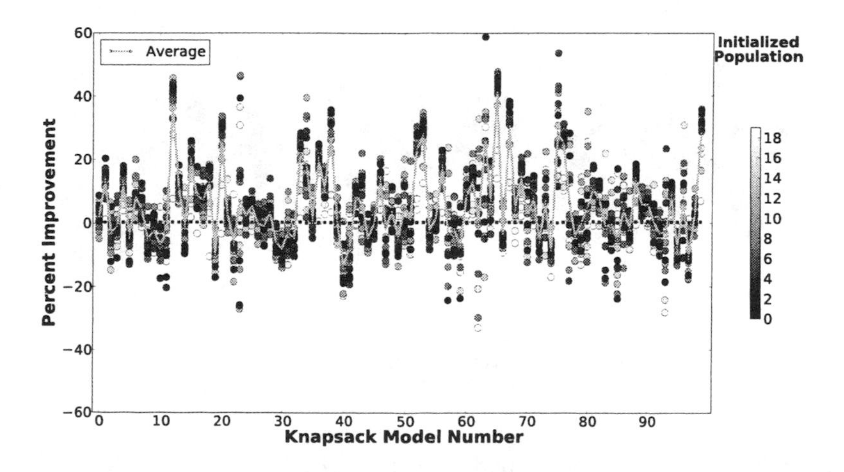

Figure 6.7 Percent difference in performance of CBDT-GA with M=100 and a profit-only utility function.

base of 500 items. Figure 6.8 shows the performance while using (6.9) with a case base of 100 items. Figure 6.10 shows the performance while using (6.9) with a case base of 500 items. In each of the percent improvement figures, the

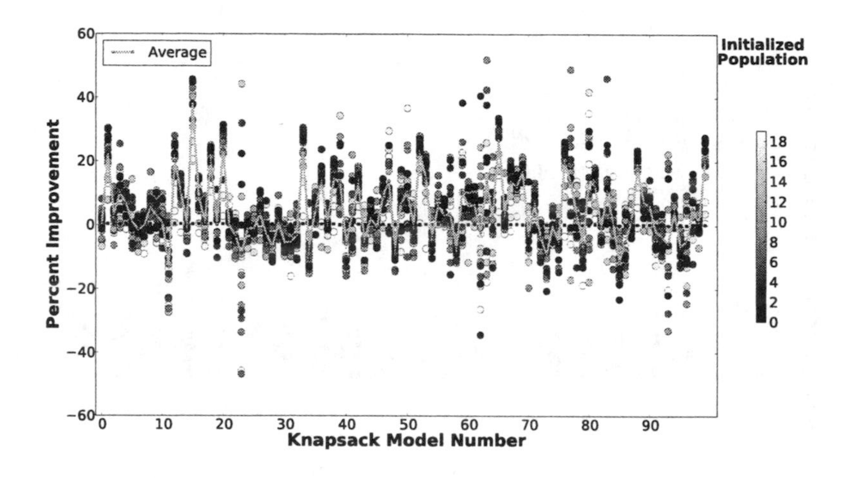

Figure 6.8 Percent difference in performance of CBDT-GA with M=100 and a weight-profit utility function.

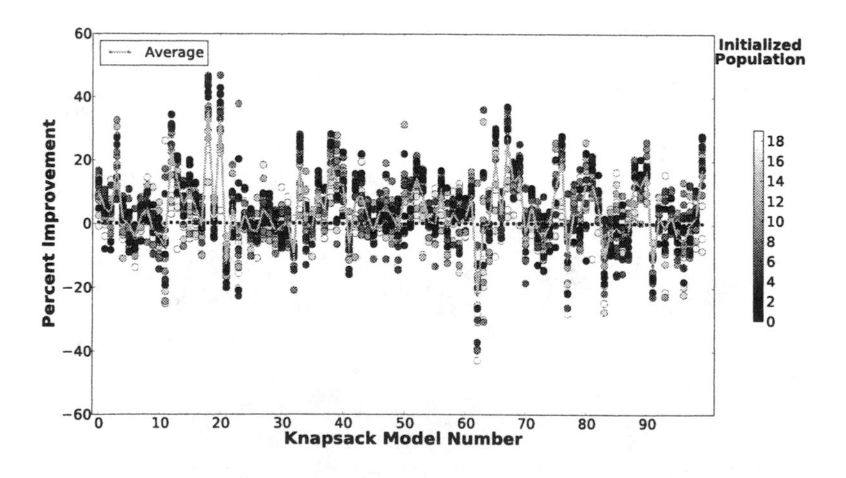

Figure 6.9 Percent difference in performance of CBDT-GA with M=500 and a profit-only utility function.

lighter circles represent larger initialized populations while the black boxes represent the base case where no individuals were initialized from the case base.

These results show some interesting aspects about the CBDT technique. First, the technique does not show improvement using a larger case base

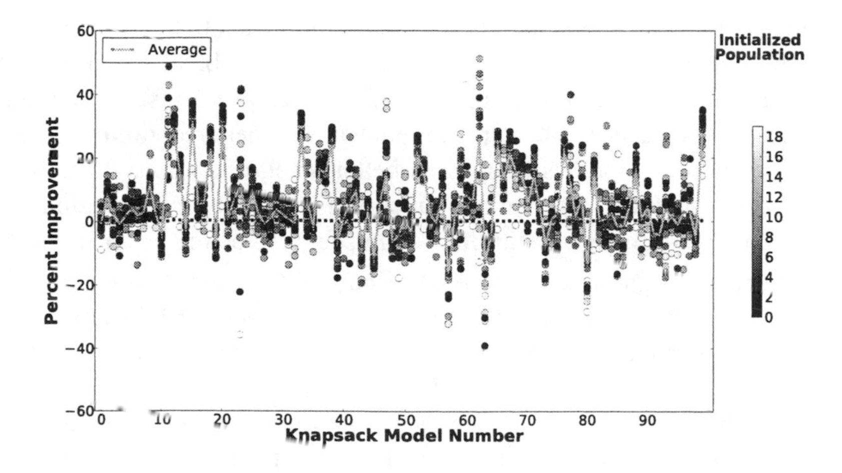

Figure 6.10 Percent difference in performance of CBDT-GA with M=500 and a weight-profit utility function.

when using the profit-only utility function. When using a larger case base, we suspect that the larger amount of experience and knowledge represented in the case base is more likely to include a better representation of the new problem. This is the situation when using the weight-profit utility function but not the utility-only function. These results suggest the significance of using the proper utility calculation when selecting the case from the case base.

To explain this trend farther, we present the optimal (or near-optimal) results of each of the knapsack models used in this analysis in Appendix E. Two significant models stand out: model 62, which has an optimal profit of 0.086188, and model 63, which has an optimal profit of 0.095681. These are both small profits and therefore difficult problems to solve by finding a small subset of items to fit in the knapsack and produce a large profit. The results of the case-based initialization show that the average improvement for model 62 with the profit-only function with 100 cases is 0.42%, with 500 items is -22.04%, the weight-profit function with 100 cases is -2.07% and with 500 cases is 26.82%. The same averages for model 63 are 26.49%, -0.33%, 13.67%, and -12.48%.

These numbers indicate that the selection of cases can have a significant impact on the results. Specifically, the initialized population will direct the genetic algorithm by offering good initial solutions. These solutions will dominate the population, and their genes will dominate during selection and reproduction. If these genes had a high utility in a previous, similar solution, they may well exist in a local optimal point in the new problem. The selection pressure exerted on the population drives the solutions into this local optimum point and not enough mutation or generations occur to break out of this local optimum before the algorithm finishes. For other problems, the initialized solutions contain genes that help move towards a better optimum. The fact that some knapsack models performed better using one utility function than another suggests that neither utility function is the best representation of the knapsack utility, although the profit-weight seems to be the better of the two.

The most significant feature of Figure 6.6 is the error bars, which represent 1 standard deviation from the mean. The standard deviation shows a significant variability in the success of the case-based initialization. The previous discussion about models 62 and 63 show why this variation exists; many problems have various success rates when applying the case-based procedure. Some problems are more easily solved then others, and the state of the case base affects how useful the initialization is to solving the problem. Sometimes, the case-based system shows significant improvement but not other times. Importantly, though, the average improvement is always positive.

Although not successful for all cases and problems, on average, the case-based approach does improve the performance of the genetic algorithm. The population initialization sometimes leads to massive improvements in the fitness of the solution after only a few generations. Some cases proved more difficult than others, and the case-based approach sometimes lead to decreased fitness values. These graphs and the trends established by them suggest some areas where the case-based system can improve to provide better overall performance.

A number of advances to this basic system are possible. Some problems show significant performance improvements by feeding in previous solutions while others do not. One improvement is the use of performance trends. For this, the implementation of the case base as a database offers many advantages. As Figure 6.3 shows, each solution in the case base is linked to a separate table that holds the fitness information. The case base can then hold many fitness values for each case and track the performance when the case is selected to initialize the optimization algorithm. If the fitness does not improve over a few runs, the decision maker can take steps in the future. When working with a genetic algorithm, these steps could include increasing the population size or altering the crossover or mutation rate.

Furthermore, tracking performance could show the decision maker that performance is not significantly improving during optimization. Given a high utility, the decision maker can assume the solutions in the case base are already near their optimal performance. In this case, the decision maker could opt to skip the optimization process or perform optimization for the new problem but with smaller population sizes, optimization run times, or other aspects that reduce the computational and time requirements of the algorithm.

Even more flexibility comes from the definition of the similarity and utility functions. Developing an understanding of the problem might help the decision maker select a different utility function when faced with a particular problem. In another potential change, the selection criteria used when selecting multiple cases from the case base simply found the best case and selected those around it. A better approach might be to find and use the N best cases, which will result in a set of solutions more properly tailored to the new problem.

6.5 Cognitive Radio Example Problem

In Chapter 3, we discussed how many different voice and data standards that are proliferating around the world for mobile phone use. Among these are the standard voice waveforms like CDMA and GSM. More exciting are the data waveforms such as Bluetooth, WiFi, Worldwide Interoperability for Microwave Acess (WiMAX), Long Term Evolution (LTE), and High Speed Packet Access (HSPA). Due to customer demand and deployment plans, wireless carriers do not support all standards, and different carriers will support better access in different areas. Mobile phones provide many different protocols including Bluetooth, WiFi, and a few different cellular standards in different bands. These are generally called *quad-mode phones* and *smart phones*. The customers usually pay a premium to buy the phones and the service plans. Already, our mobile phones switch between roaming and nonroaming carriers and change between a small subset of possible protocols. These selections are performed by the radio itself, sometimes with specific rules provided by the user, such as to save cost by never roaming.

Besides differences in service provided by the carriers, different standards offer various levels of service quality. Among the big differentiators are the range of the data rates available and the latency of the network. WiMAX, for example, has a higher latency than LTE, making LTE more applicable to voice communications.

We want our radios to be capable of all of these waveforms and modes of service and to automatically select the service based on some quality of service metrics. We have so far discussed technical quality of service measures, such as speed, error rates, and power consumption. Another very important input to a consumer's decision-making method is the cost of the service. This concept has been addressed in many of the papers published in the IEEE DySPAN symposia, including [8].

Table 6.1
Comparison of Mobile Data Service Technology

Technology	Throughput	Latency	Power	Cost
CDMA / 1×RTT	144 kbps	High	Low	Med.
CDMA / EV-DO	3.1 Mbps	Med.	Med.	Med.
GSM / GPRS	114 kbps	High	Low	Low
GSM / EDGE	236.8 kbps	Med.	Low	Low
UMTS / HSPA	14.4 Mbps	Low	Med.	Med.
802.11b	11 Mbps	Med.	Med.	Low
802.11g	54 Mbps	Med.	Med.	Low
WiMAX	75 Mbps	Med.	High	High
LTE 2×2	172.8 Mbps	Low	High	High
LTE 4×4	326.4 Mbps	Low	High	High

In our scenario to establish a decision-making model for cognitive radios, we have to make a few assumptions. First, let us assume that there is a practical method of micropayments established between all mobile carriers such that the cognitive radio can purchase service for a small amount of time. This scenario does not assume or analyze anything related to spectrum auctions or price bargaining [9]; instead, the service provider offers the consumer a specific price for using their service over a specified length of time. We also assume that not all service providers or services are available in all geographic locations.

Table 6.1 reflects some of the information the cognitive radio can use to make its decisions for which service is most suitable to an application. The throughput values listed in the table are the published theoretical maxima for the technology, although we know that no one subscriber unit could achieve these values. We list them in this way as a way to allow comparison between the different waveforms. The latency, power, and cost are listed qualitatively as low, medium, or high for comparison purposes. Power can depend on the handset in use, and cost and latency depend greatly on the service provider.

With no previous experience, a radio can use Table 6.1 to make a decision on what technology to use for a given application's needs. Here are some examples: use LTE for streaming live video, WiMAX for emailing a proposal, or any of the lowest power and cost services to send a grocery list. However, not all services will offer every technology at a given location, and different services will exhibit different performance. The differences could be due to poor network management, heavy user capacity, or geographic effects on the signal propagation. These facts can not be known in advance from just the information about the technology. This kind of knowledge must be

Service Provider Case-Base								
Service	Technology	Location	Time	Throughput	Latency	Power	Dropped Calls	Blocked Calls

Figure 6.11 Case representation of the cognitive radio case-based decision maker for service selection.

taught or learned from experience over time. Experience can show that a given service is highly time- and location-dependent. Peak traffic times and poor signal quality can lead to more dropped calls, more blocked calls, and increased latency in the network.

We can use a case-based decision making engine to solve these problems. The case base can learn about service utility, such as the probability of a dropped connection or slow service. The case base would have to be programmed based on observable network performance. The case base would update its information for a given network connection, including the network speed and latency and experience with lost connections. All of this information would have to be tagged with the given time and location. The case base would look like Figure 6.11.

Given the knowledge represented in the case base, the decision maker must be able to use it in order to determine which service to use for its current application. Using the similarity and utility function analysis we developed, we must now extend these functions for the given problem.

The similarity is determined by the environment. The environment consists of the time of day, day of the week, location, and available service providers. This is another multidimensional problem, so the problem is to figure out how to combine the information to produce an understanding of the similarity. Service at the same time of day on a weekday in any given city might have similar problems due to heavy network use. At other times, usage patterns can be specific to certain locations, such as the difference between urban and rural roads.

In our academic exercise here, we have to recognize that this kind of similarity analysis is very difficult without surveying and modeling the influences of location and time. To simplify the scenario, we will break the concepts down to simple measurements of location in the form of longitude and latitude and time in the form of the time of day and day of the week. Further investigation could help tie in concepts like location diversity, such as the difference between rural, suburban, and urban areas. The similarity function of (6.10) compares the time and location by calculating an error between the current location, $d(p)$, and the location of the case, q, in the case base, $d(q)$. The time of day is calculated similarly as an error between the current time, $t(p)$, and time in the case base, $t(q)$. The day of the week is handled differently in that the similarity is 1 if the day of the week, $D(p)$, is the same as the day of the week in the case base, $D(q)$. Otherwise, it is 0. A more complicated model could have a partial similarity that equates to 0.5 if D and $D(q)$ are both weekdays or both weekends.

$$\begin{aligned} s(p,q) &= 1 - \tfrac{1}{3}\left(\tfrac{|d(p)-d(q)|}{d(q)} + \tfrac{|t(p)-t(q)|}{t(q)} + (1 - D_\Delta)\right) \\ D_\Delta &= \begin{cases} 1 & D(p) = D(q) \\ 0 & D(p) \neq D(q) \end{cases} \end{aligned} \tag{6.10}$$

The utility is based on the observables of the network service and is maximized by higher throughput, lower BER, latency, dropped calls, and blocked calls. This is related to the multiobjective problem where these different values cannot be directly compared or combined due to differences in scale and importance to the situation. On the other hand, this problem is constrained to known services, so unlike the general cognitive radio system, we have a known set of parameters and values. While knowledge is being built up by the case base, the system selects services based on the required QoS, so it has already primed the decision making with known desired values. We will ignore the BER and treat the throughput as an indirect measure of this value. We can use this fact to build the utility measure of future systems. For each service q in the case base, the cognitive radio measures the performance, such as average throughput, $\hat{R(q)}$, and latency, $\hat{L(q)}$ during the connection. These values are directly comparable to the advertised service of the case, $R_s(q)$ and $L_S(q)$ for the throughput and latency, respectively. The radio would also keep track of the number of dropped calls, $C_D(q)$, and blocked calls, $C_B(q)$, to calculate a percentage of each value occurring over the total number of connections made or attempted, C. The utility function in (6.11) can be built from the knowledge and assumptions. This function sums the error between the actual service and the proposed service as well as the percentage of dropped and blocked calls. The closer to 1, the better the utility of the service.

$$u(q) = 1 - \frac{1}{4}\left(\frac{|\hat{R} - R_s(q)|}{R_s(q)} + \frac{|\hat{L} - L_s(q)|}{L_s(q)} + \frac{C_B}{C(q)} + \frac{C_D}{C(q)}\right) \tag{6.11}$$

An aspect of the case-based approach that should become obvious as we discussed this example is the ability to use knowledge developed from other users to augment our own. Our example has discussed the use of past information at different times and different locations. To individually build up enough information to make good, informed decisions could take a long time. If, however, other users who have also been developing similar information can share this, then the aggregate information built among all radios is more useful. There is of course security issues with this, such as poisoning the case base.

6.6 Conclusion

In this chapter, we introduced the concept of using feedback in the optimization process to aid future optimizations. We used the concept of case-based decision theory as the mechanism for producing the feedback. When one problem is optimized, the solution is fed back into a case base that stores the solution, results, and problem. When a new problem is received, the decision maker looks for similar problems in the case base that exhibit a high utility. The previous solutions stored in the case base are then fed to the optimization process. We examined how this technique is applied to the cognitive engine to build up knowledge and learn from experience.

To demonstrate the concept, we employed the case-based learning system to the simple knapsack problem. The results showed an overall performance benefit from the implementation. They also showed that there are still significant gains to be made by further studying this technique as well as a number of parameters that can be adjusted. The case base size, as well as the number of cases used to seed the optimization algorithm, can be changed for different results. More importantly, the results in this chapter showed that the definition of the utility function can have a great impact on the performance. We suggest that the similarity function would also produce a large performance difference. With the knapsack problem, we only present one similarity function that made sense for this problem, but for more complicated problems, different similarity rankings may be possible and so must be studied.

There are also advances from this concept beyond these algorithm adjustments. As we discuss at the end of this chapter, this method introduces other aspects of learning, such as using performance trends to build up confidence or alter behavior. This concept offers a number of significant advantages to online optimization processes, especially when strict time limits must be imposed. Because of this, case-based learning provides significant potential for use with the cognitive engine where problems require solutions quickly as situations and environments change.

We will show the results of using case-based learning in the cognitive engine in Chapter 8. First, however, we discuss a bit more theory and design of the cognitive engine. Chapter 7 covers some important topics of looking at the cognitive engine within a network of radio nodes. The cognitive engine develops a waveform that all nodes on the network will then need to use. The next chapter discusses potential methods for coordinating the networks and distributing the information among the nodes. We also provide a brief discussion of the concept of using distributed algorithms to improve performance and decision making.

References

[1] T. W. Rondeau, B. Le, D. Maldonado, D. Scaperoth, A. B. MacKenzie, and C. W. Bostian, "Optimization, Learning, and Decision Making in a Cognitive Engine," *Software Defined Radio Forum Technical Conference*, 2006.

[2] D. E. Goldberg, *Genetic Algorithms in Search, Optimization, and Machine Learning*, Reading, MA: Addison-Wesley, 1989.

[3] C. L. Ramsey and J. J. Grefenstette, "Case-Based Initialization of Genetic Algorithms," *Proc. Fifth Int. Conf. Genetic Algorithms*, 1993, pp. 84 – 91.

[4] T. R. Newman, R. Rajbanshi, A. M. Wyglinski, J. B. Evans, and G. J. Minden, "Population Adaptation for Genetic Algorithm-Based Cognitive Radios," *IEEE Proc. Cognitive Radio Oriented Wireless Networks and Communications*, Aug. 2007.

[5] I. Gilboa and D. Schmeidler, *A Theory of Case-Based Decisions*, Cambridge: Cambridge University Press, 2001.

[6] J. Kolodner, *Case-Based Reasoning*, San Mateo, CA: Morgan Kaufmann Pub., 1993.

[7] C. J. Rieser, "Biologically Inspired Cognitive Radio Engine Model Utilizing Distributed Genetic Algorithms for Secure and Robust Wireless Communications and Networking," Ph.D. diss., Virginia Tech, 2004.

[8] S. Ball, A. Ferguson, and T. W. Rondeau, "Consumer Applications of Cognitive Radio Defined Networks,"*IEEE Sym. New Frontiers in Dynamic Spectrum Access Networks (DySPAN)*, 2005, pp. 518 – 525.

[9] L. Doyle and T. Forde, "Towards a Fluid Spectrum Market for Exclusive Usage Rights," *IEEE Sym. New Frontiers in Dynamic Spectrum Access Networks (DySPAN)*, 2007, pp. 620 – 632.

7

Cognitive Radio Networking and Rendezvous

A final challenge to enable the cognitive radio system's basic functionality is the ability to transmit the cognitive engine's information and solutions among the nodes operating on the network. We have analyzed the cognitive engine from the internalized view of optimizing a waveform. The next item to address is the method by which a cognitive radio acts as part of a network.

A cognitive network is more than a network of cognitive radios. It exhibits distributed intelligence by configuring and adapting individual nodes to meet a dynamic set of network-level goals. The behavior responds to goals of the radio users but is not controlled by any individual node in the network. Within the networks constraints, they configure themselves to best meet their own users' needs.

At the time we prepared this book for publication, cognitive networking was in its infancy. Practical developments consisted largely of a few nodes controlled by a single cognitive engine or else were limited to DSA applications in which nodes share a central database. There were lots of speculative proposals for optimization at the network level, emergent behavior or swarm theory as models for cognitive networking, and so forth. We debated omitting this chapter, but we decided to include it as a summary of where the field stood in late 2008.

The most basic implementation is an autocratic method of cognitive radio network development. In this method, one radio develops a waveform and pushes it out to the nodes for them to use. The biggest challenge to enabling this mechanism is the need to establish communications among all nodes before the radio can communicate the new waveform. Developing this theme further, the autocratic method falls short of realizing the full potential of a cognitive radio network. When one cognitive radio develops a waveform, it has developed it to optimize its internal goals for its perceived channel conditions. The other radios on the network may not share these conditions, and so one radio's optimized waveform may not be the same as another radio's. We briefly discuss this with respect to the literature of game theory and cognitive networks that have been working on this issue.

A further enhancement to the cognitive radio design is not only to distribute the waveform information, but also the use of the network nodes to enhance the optimization process. Each cognitive radio in a network has the ability to cooperatively optimize through the use of distributed and parallel processing. We end this chapter by addressing some of the very basic aspects of these techniques with respect to enhancing the genetic algorithm.

This chapter addresses cognitive radio networks, and each of the topics discussed here are full research endeavors on their own. Our aim in this chapter is to develop the basic system to deliver waveforms across a network and present the research areas involved and the advances this topic has to offer.

7.1 Waveform Distribution and Rendezvous

The simplest approach to enabling communications among cognitive radio nodes is through a static control channel. In the first scenario under this model, the radios in a network are currently in communication with each other. One or more of the cognitive radios then develops a new waveform that improves communications. Under this condition, the radio can simply pass the new waveform to the radio nodes using the current channel. This method is a form of *in-band signaling* and can use a different logical control channel over the same physical channel to send configuration information. This type of control information is commonly used in home networking systems like IEEE 802.11, where connection and configuration data use the same frequency channel but a simpler, more robust modulation scheme.

In another scenario, the radios in the network are not able to communicate with each other due to degradation in the channel such as increased interference or environmental changes. In this case, the radios would have to use another physical channel known to all radios to reestablish communications. This is a form of *out-of-band signaling*, where the radios use a separate physical channel to communicate control information. This concept is commonly used in cellular communications systems. The control channel is defined to use simple, robust waveforms on which all nodes are capable of communicating. In the worst case, if the cognitive radio nodes lose communications, they can revert to the control channel and wait for the new waveform information and then reestablish communications. The control channel is also used to begin communications when a node wants to join a network that might be using any waveform or any frequency. The control channel allows the new node a way to communicate with the network and initialize communications. This concept is often referred to as *rendezvous*: the method by which a radio hails and enters a network.

Ideas and implementations for rendezvous are receiving a lot of attention for cognitive radio network coordination and many papers of recent dynamic spectrum and cognitive radio conferences discuss this Both the 2005 and 2007 IEEE DySPAN conference proceedings contain a number of such papers; for example, see [1, 2, 3].

Static control channels, while easily implemented, are problematic because they are easily jammed and rendered useless. More innovative ideas involve dynamic control channels, which still require coordination among the nodes to determine where the control channel is. A few proposals have been shown that remove the control channel from the rendezvous model and instead use physical layer descriptors to identify radios and enable rendezvous. Sutton et al. [4] shows the use of embedded cyclostationary signatures in OFDM-based systems that can identify a network and coordinate access. Because the signature is embedded in each OFDM symbol transmitted, the system does not need to transmit particular frames or switch channels to enable the network identification and coordination. Horine [5] proposes a technique to search for clear channels, transmit a beaconing signal, and wait for a response while other radios scan for the particular beacon. This concept is similar to Bluetooth's inquiry and connection states. The beacon is shaped in frequency to identify the node or network. Unfortunately, since the detection is based on FFT amplitude, there is no offered explanation of how the approach will work in multipath or fading channels.

There is significant interest and work progressing in cognitive radio rendezvous. Because of its simplicity of implementation and the currently available SDR capabilities, we have implemented the static control model to enable our experiments with the cognitive engine. The cognitive engine first tries to contact the other radio nodes on the current channel; if they do not respond, the radio reverts to a known control channel and waveform where the other nodes, having likewise lost their connection, will wait. The new waveform information is then transmitted to the nodes after which they reconnect using the new settings.

7.2 Cognitive Radio Networks

The static control channel model where waveforms are pushed to the radios is currently used in the our cognitive engine implementation for lack of a better solution. This method also ignores the possibility that a waveform created by one radio does not work for another radio. In a heterogeneous network, some nodes may be incapable of using the particular waveform. Even if all nodes are capable of using the specified waveform, other aspects of the waveform may perform badly for certain nodes. A major issue often discussed is the hidden node problem. A radio that is unseen by the cognitive radio designing the waveform might be in close proximity to another node on the network. The new waveform, while good for the designing node, causes interference to the other nodes.

Research in cognitive networks, such as the work by Thomas et al. [6, 7], attempts to address this issue by looking at end-to-end performance. From this perspective, the cognitive network uses objective functions that optimize with respect to the network performance. In [6], they use a game theory approach to optimize an ad hoc network with respect to power and channel control. Game theory has been widely studied for wireless network optimization to look for optimal states for all nodes, or a *Nash equilibrium*. Neel provides an extensive discussion and analysis of game theory for cognitive radio [8].

7.3 Distributed AI

Another benefit from looking at the whole network instead of single node adaptation is the advantage of the available processing power capabilities of each node. Parallel processing has often been used advantageously in computer science, and with the move towards multicore processors, it is likely a subject that will continue to receive attention. Some algorithms have shown themselves to be easily separable for processing portions on different processors, and genetic algorithms are among these. Goldberg cites many methods that take advantage of the populations of a GA in a distributed sense [9].

A particularly popular technique is to split the population among different processing elements to create "islands" of populations. Each population is independently optimized, and the populations of optimized chromosomes are combined, compared, and a winner is selected. Populations can also migrate between islands to share genes. Alba provides a background on parallel genetic algorithms by applying them to find appropriate error-correcting codes and antenna placement in a radio network [10]; these applications are particularly interesting in light of our work. Other literature on this topic looks more closely at the mechanisms at work in the migration and populations [11] as well as other types of parallelization of GAs [12, 13].

When applying a parallel genetic algorithm to an online learning system such as a cognitive radio, there are many questions that need to be addressed.

The parallel GAs have some form of migration, or sharing, of population members to perform the global analysis of the results to find a solution. The implementation of migration should be designed to consider the required network overhead. Another issue is that cognitive radio networks are dynamic where nodes can come and go at random. Most parallel GAs are studied under the assumption that the network of processing elements was established for this task. Instead, a parallel GA in a cognitive radio network performs the parallelization as a secondary process to its normal communications. The algorithm must be implemented with respect to the dynamics of the network and robust against the loss of processing nodes. Distributed AI offers significant potential to improve the global solutions and reduces the time and power required by any individual node, but these are some of the issues around which such a distributed system must be implemented.

7.4 Conclusions

While giving few answers, we have discussed some important considerations in the future development and deployment of cognitive radios in this chapter. A network of cognitive radios must include methods by which to transfer waveforms among all nodes as well as take into consideration the needs of other nodes when designing new waveforms. The networking aspect itself opens up the potential to use distributed and parallel processing to enhance cognition in the network. In any case, consideration must be given to the overhead required on the network to transfer the information related to the cognitive radio performance and network maintenance.

For the practical purposes of the experiments in the next chapter, the method used for node control is a simple push method from one node to the others when it develops a new waveform. Furthermore, a default waveform provides a fallback channel and modulation for the nodes to use if communications is lost.

References

[1] J. Perez-Romero, O. Sallent, R. Agusti, and L. Giupponi, "A Novel On-Demand Cognitive Pilot Channel Enabling Dynamic Spectrum Allocation," *IEEE Proc. DySPAN*, 2007, pp. 46 – 54.

[2] C. Cordeiro and K. Challapali, "C-MAC: A Cognitive MAC Protocol for Multi-Channel Wireless Networks," *IEEE Proc. DySPAN*, 2007, pp. 147 – 157.

[3] J. Zhao, H. Zheng, and G. Yang, "Distributed Coordination in Dynamic Spectrum Allocation Networks," *IEEE Proc. DySPAN*, 2005, pp. 259 – 268.

[4] P. D. Sutton, K. E. Nolan, and L. E. Doyle, "Cyclostationary Signatures for Rendezvous in OFDM-Based Dynamic Spectrum Access Networks," *IEEE Proc. DySPAN*, 2007, pp. 220 – 231.

[5] B. Horine and D. Turgut, "Link Rendezvous Protocol for Cognitive Radio Networks," *IEEE Proc. DySPAN*, 2007, pp. 444 – 447.

[6] R. W. Thomas, R. S. Komali, A. B. MacKenzie, and L. A. DaSilva, "Joint Power and Channel Minimization in Topology Control: A Cognitive Network Approach," *IEEE ICC*, Jun. 2007, pp. 6538 – 6543.

[7] R. W. Thomas, D. H. Friend, L. A. DaSilva, and A. B. MacKenzie, "Cognitive Networks: Adaptation and Learning to Achieve End-to-End Performance Objectives," *IEEE Communications Magazine*, Vol. 44, No. 12, pp. 51 – 57, Dec. 2006.

[8] J. Neel, "Analysis and Design of Cognitive Radio Networks and Distributed Radio Resource Management Algorithms," Ph.D. diss., Virginia Tech, 2007.

[9] D. E. Goldberg, *Genetic Algorithms in Search, Optimization, and Machine Learning*, Reading, MA: Addison-Wesley, 1989.

[10] E. Alba and J. M. Troya, "A Survey of Parallel Distributed Genetic Algorithms," *Complexity*, Vol. 4, No. 4, pp. 31 – 52, 1999.

[11] W.-Y. Lin, T.-P. Hong, and S.-M. Liu, "On Adapting Migration Parameters for Multi-Population Genetic Algorithms," *IEEE Proc. Systems, Man and Cybernetics*, 2004, pp. 5731 – 5735.

[12] J. P. Cohoon, W. N. Martin, and D. S. Richards, "Punctuated Equilibria: A Parallel Genetic Algorithm," *Proc. Int. Conf. Genetic Algorithms*, 1987, pp. 148 – 154.

[13] L. Chambers, *Practical Handbook of Genetic Algorithms: New Frontiers*, Boca Raton, FL: CRC Press, 1995.

8

Example Cognitive Engine

Over the past few chapters, we have been building up the design of our cognitive engine. We now present an implementation of the cognitive engine along with examples in both simulation and over-the-air experiments. We develop the experiments in four parts: a simple use of the genetic algorithm to design waveforms in the simulation environment, addition of the interferers to the example, the use of the case-based decision theory feedback mechanism, and finally the application of the cognitive engine over the air.

One of the major problems we have to address when presenting the results is the lack of a definitive mechanism to compare results. Unlike the knapsack problems presented previously where only one metric represents how successful the optimization was, the results here are multidimensional. Furthermore, for the complex environments, many potential solutions exist to produce desired behavior, and the random, nontractable behavior of the genetic algorithm may yield different waveforms giving the same results. To work through this, we build up the simulation analysis slowly, introducing new objectives each time for specific purposes and showing the behavior of the genetic algorithm by plotting the performance of each objective over the generations. The trends exhibited in these graphs will indicate the response of the optimization across the different objectives. As the number of objectives increases, the distribution of the solutions along the Pareto front becomes more difficult to see. Therefore, the early trend graphs are designed to show the success in the optimization method while later results will be measured more empirically. Finally, all graphs of the same objective functions are plotted using the same range so each can be compared to the others.

8.1 Functional System Design

While in the previous chapters, we discussed the theory behind the system design and explained how the different systems work together, in this section, we review the full design of the cognitive engine package. Figure 2.3 showed the generic cognitive engine system featuring its major components. Figure 8.1 shows the specific implementation of the cognitive engine for the simulations. Note that the verification system has been removed from these simple experiments. Since we are focusing on the optimization of waveforms, under the simulation environment, we have no need to add policy decisions. Later, we will show some experiments done under specific regulatory constraints and how the cognitive engine responds to these conditions.

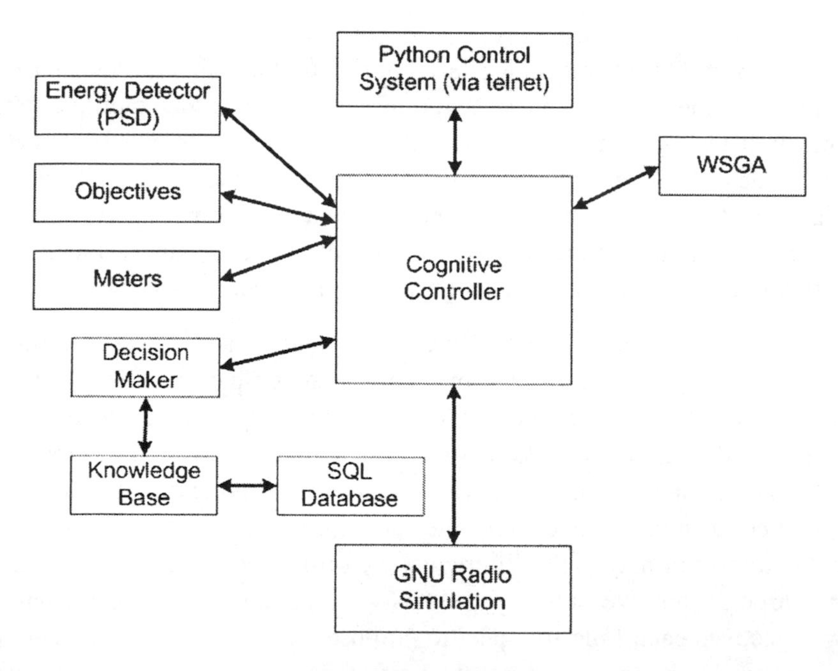

Figure 8.1 CWT's cognitive engine implementation for the simulated experiments.

First, each of the sensor components is instantiated; these include the meters, the objectives, and the PSD sensors. The meters sensor provides information about the performance of the radio by "reading the meters." The objectives sensor provides information gathered about the QoS requirements of the user and the application, and it indicates the objectives to use in the analysis. The PSD sensor provides information about the external interference environment, or power spectral density of the spectrum. The

XML representation of the information from all of these meters is provided in Appendix D.

Next, the optimization component needs to be instantiated. Like the sensors, this is a state machine process that listens for a connection and a request to process data from the cognitive controller. When this happens, the cognitive controller passes a number of items to the optimization process which is either required for processing or adds information to help in the optimization. In this case, the optimization routine is the WSGA, which requires the set of algorithm parameters, the problem definition, and the chromosome definition. The parameters are the operating parameters such as the population size, the crossover and mutation rate, and the termination conditions such as the maximum number of generations. The problem definition describes the optimization evaluation. For the WSGA, the problem statement is the set of objective functions to use, such as BER or throughput (see Chapter 4), and the weight of each objective. Finally, the chromosome definition describes the radio hardware. The chromosome is a binary vector that represents the search space; in the WSGA, the search space is the set of radio knobs. The DTD and XML description of the waveform (see Appendix D) provide the mechanism to map between the chromosome and the radio knobs. This process is discussed in more detail in Chapter 5.

Three extra pieces of information can also be sent to the optimization process. The first are previous solutions used to seed the population from the case-based decision maker. Also, the radio environment map can be sent to the optimizer to help it understand the interference environment. This can then assist with the waveform design like the center frequency and bandwidth of the signal. The cognitive controller can also send the results of the meters sensor describing the measured noise power and path loss, which are useful in determining the parameters such as required transmit power and modulation type for the current conditions. The optimization process can run without any of these three pieces of information, but each is meant to help inform and guide the optimization process for better performance.

The radio framework then needs to be instantiated. The framework, as described in Chapter 2, provides the translation between the cognitive engine's representation of the waveform and the radio platform. In this case, it parses the XML waveform representation and creates a GNU Radio flow graph.

When instantiated, the cognitive engine associates itself with the sensors, optimization process, and radio framework that are part of the cognitive engine. A simple XML file tells the cognitive controller what systems are present and how to contact them. Each sensor is first contacted and stored as a component associated by a specific name for the sensor, so there can be any number of sensors attached at any given time. Only one optimization and radio framework are currently supported at a time. The cognitive engine can also describe the policy engine interface, if one exists, and it configures the user interface. Currently, the decision maker and case-based systems are integrated with the cognitive controller, but these should be separated to allow different implementations. When this happens, the decision maker, too, will associate with the cognitive controller in this standard way.

When a component is associated, the cognitive controller requests the XML and DTD files used by the component to describe its information. These files are used to build the representation in the case base and store the results of each component, like the optimization results or the sensor data. It is exchange of information that necessitates the instantiation of the components before the controller.

Below, we show a simple Python script that acts as the user interface to control the cognitive engine. The command structure is simple:

<component>:[name]:<command> [extra parameters]

The *component* is one of the major components, such as "sensor" or "optimizer." The *name* is only required for the sensors to describe which sensor the command is to be passed to, such as "meters" or "objectives." The *command* describes action requested from the component, such as "collect" to request data from the sensor or "optimize" to tell the optimization process to run. The final optional *extra parameters* can be additional information to a component for its processing, such as a frequency range for the PSD sensor to scan.

The control script performs the actions listed here:

1. Run the radio system;
2. Collect the meters and PSD;
3. Collect the objectives;
4. Search for previous solutions in the case base;
5. Run the optimization process given the problem definition, the chromosome definition, the GA parameters, the radio environment map and meters, and the previous solutions;
6. Receive the optimized waveform and simulated results and store these in the case base;
7. Run the new waveform on the radio in the environment;
8. Collect the new meters and compare the performance.

8.2 Simple Simulations

The simple simulation environment uses an AWGN channel with a free-space path loss shown in (8.1) where $n = 2$, $G_T = G_R = 0$, c is the speed of light in meters per second, d is the distance between transmitter and receiver in meters, P_T is the transmitted power in dBm, and P_R is the received signal power in dBm. The path loss is implemented by using a frequency of 780 MHz. We chose this frequency because of the significant interest in this spectrum and the FCC 700 MHz spectrum auction that took place in early 2008. The distance was set for this frequency to provide a path loss of about -22 dB, or about 1 wavelength.

$$P_R = P_T + G_T + G_R - n10log_{10}\left(\frac{4\pi df}{c}\right) \tag{8.1}$$

The simulation ran 100 kB, or 8×10^5, bits through the transmit-receive chain, which will provide adequate representation of the BER where 1×10^{-6} is considered the lowest threshold and most of the simulations were designed for higher BER than this. The objective function calculation was the simple lin-log of (5.2).

8.2.1 BER-only

In the following tables, we abbreviate the columns for bit error rate (BER), signal to interference plus noise ratio (SINR), throughput (Thr.), bandwidth (BW), spectral efficiency (Spec. Eff.), interference power (Int.), power consumption (Pwr.), and computational power (Comp.). Table 8.1 shows these values for the BER-only test. The results of the test are listed in Table 8.2.

Table 8.1
Objectives: BER-Only Test

BER	SINR	Thr.	BW	Spec. Eff.	Int.	Pwr.	Comp.
1.00	N/A	N/A	N/A	N/A	N/A	0.00	N/A

Figure 8.2 shows the performance of the BER and power objectives over the generations of the genetic algorithm. These plots show the expected behavior of the objectives as the BER objective is optimized. The power objective, while calculated, had a weight of 0, so it did not factor into the preference of the algorithm. We use the calculation here to show how the two objectives are traded off under the performance criteria. The BER curve has a steady negative slope for the best performing individuals while the power increases as a direct result of minimizing the BER.

Table 8.2
Waveform Settings and Results: BER-Only Test

Knob	Settings
Modulation	BPSK
Transmit power (dBm)	19.08
Symbol rate (sps)	0.250
Pulse shaping	RRC, 0.86
Normalized frequency	−0.620
Packet size	171
Meters	**Sim. Result**
BER	1.18×10^{-7}
SINR (dB)	N/A
Spec. Eff. (bps/Hz)	N/A
BW (Hz)	N/A
Throughput (bps)	N/A
Interference (dBm)	N/A
Power (dBm)	19.08
Computation (ticks)	N/A
Obs. BER	0

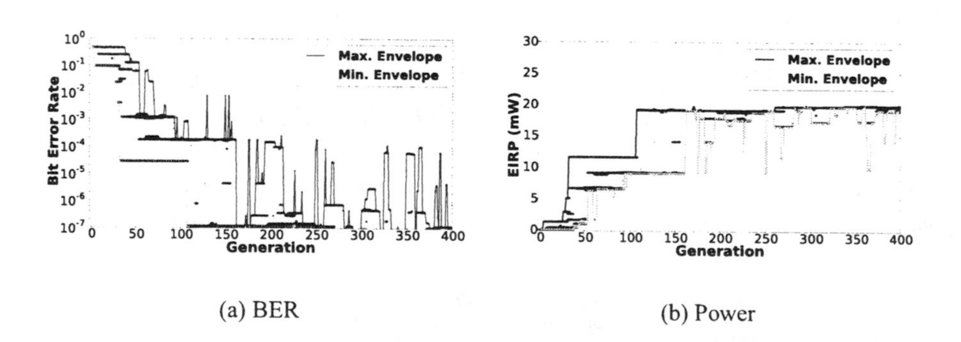

(a) BER

(b) Power

Figure 8.2 Performance curves for BER-only test with plots for objectives (a) BER and (b) power.

The graphs in Figure 8.2 are simple representations of the graphs we will be discussing in more complicated, multiobjective scenarios. Even in this simple simulation, the graphs show the random behavior of the genetic algorithm that makes them difficult to read. The BER plot shows that even in later generations, some individuals exhibit large BER due to mutations or bad crossover. As we introduce more objectives into the experiments, different optimization forces will cause more complex responses by the GA that make these curves even more difficult to follow.

8.2.2 BER and Power (1)

For a more interesting example than the results just presented, in this simulation, we add the power minimization objective to the algorithm and lower the BER objective weight as represented in Table 8.3. The algorithm should now produce a waveform that produces a low BER while not driving the power to the maximum, and indeed, Table 8.4 shows this exact trend. The BER is higher than in the previous example and the power is reduced.

Table 8.3
Objectives: BER and Power Test (1)

BER	SINR	Thr.	BW	Spec. Eff.	Int.	Pwr.	Comp.
0.75	N/A	N/A	N/A	N/A	N/A	0.50	N/A

Table 8.4
Waveform Settings and Results BER and Power Test (1)

Knob	Settings
Modulation	BPSK
Transmit power (dBm)	12.88
Symbol rate (sps)	0.250
Pulse shaping	RRC, 0.22
Normalized frequency	−0.807
Packet size	101
Meters	**Sim. Result**
BER	3.28×10^{-2}
SINR (dB)	N/A
Spec. Eff. (bps/Hz)	N/A
BW (Hz)	N/A
Throughput (bps)	N/A
Interference (dBm)	N/A
Power (dBm)	12.88
Computation (ticks)	N/A
Obs. BER	0

One of the more telling aspects of this result is the value of the BER result of the cognitive engine versus the observed result of using the waveform on the radio. The simulated BER looks a bit large and suggests that the algorithm did not perform as well as expected; from the objective function settings, the waveform should more heavily weigh minimizing the BER to minimizing the power. However, Figure 8.3 shows why this happened. The BER plots never managed to produce a BER value lower than about 1.0×10^{-2}, even when the power approached the maximum of 20 mW. This is a result often observed when running the algorithm due to the uncertainty of the meters sensor. The objective for setting the BER, as discussed in Chapter 4, relies on the approximation of the noise floor and the path loss. As the results of Appendix F show, the estimation of the SNR can be off by a few decibels, plus or minus. At the power levels the algorithm is dealing with, a few decibels can greatly affect the simulated performance, and so a bad estimate can produce results such as these. However, the actual observed result of using the waveform showed a much better BER than the simulated performance, indicating that this was actually a good waveform choice that produced the proper balance of power to BER. Furthermore, this indicates a robustness under uncertainty in the system performance.

It does not matter in this case that the information received from the sensor was incorrect; the algorithm found the proper balance. We have previously discussed the concept of overoptimization. An error in the meter's representation could lead to an artificially increased transmit power to the detriment of other nearby radios as well as its own power consumption.

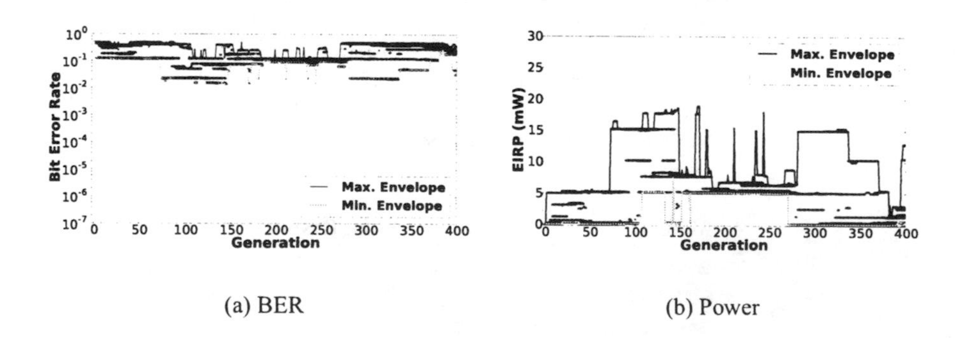

(a) BER (b) Power

Figure 8.3 Performance curves for BER and power test (1). As (a) BER and (b) power are competing objectives, the algorithm must balance them.

8.2.3 BER and Power (2)

To study the behavior a bit further, an interesting question to address is the effect changes in the weights have on the performance of the cognitive engine. Using Table 8.5, we dropped the BER weight to 0.5 so that the optimization process has no preference. Of course, only the relative value of the weights matter, so this test would work equally well by setting both weights to 1.0. The results in Table 8.6 show one extreme of what can happen in this situation. The power is 0.28 dBm, almost the minimum power available on the system, but the BER value is terribly large. This results from the lack of preference and improper shaping of the objective space. Without a real preference relationship established, one extreme or another is likely to take over. Other runs under these conditions showed very small BER and large power values. However, we use this example to illustrate the performance of the GA in Figure 8.4, which shows interesting performance trends.

Table 8.5
Objectives: BER and Power Test (2)

BER	SINR	Thr.	BW	Spec. Eff.	Int.	Pwr.	Comp.
0.50	N/A	N/A	N/A	N/A	N/A	0.50	N/A

Table 8.6
Waveform Settings and Results: BER and Power Test (2)

Knob	Settings
Modulation	BPSK
Transmit power (dBm)	0.28
Symbol rate (sps)	0.750
Pulse shaping	RRC, 0.22
Normalized frequency	−0.647
Packet size	164
Meters	**Sim. Result**
BER	3.41×10^{-1}
SINR (dB)	N/A
Spec. Eff. (bps/Hz)	N/A
BW (Hz)	N/A
Throughput (bps)	N/A
Interference (dBm)	N/A
Power (dBm)	0.28
Computation (ticks)	N/A
Obs. BER	3.995×10^{-1}

Figure 8.4 shows what happens when two objectives are equally traded off. The curve in the middle generations shows that during the first hundred generations the algorithm is looking for a small BER and large power value, but then more highly fit individuals take over the population and push it in the other direction. Again, because there is no selection pressure influencing the

population away from either extreme, both may be equally fit. We addressed this concern briefly in Chapter 4 in discussing different utility functions. Replacing the linear summation of weights used in these tests to one of the utility function might show improvement under these conditions.

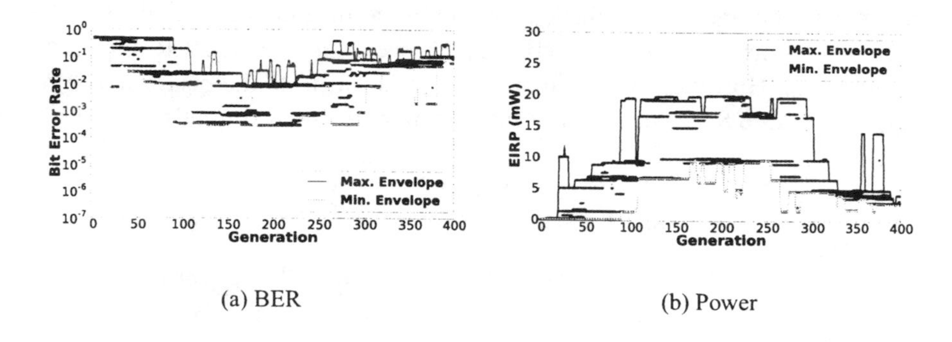

(a) BER

(b) Power

Figure 8.4 Performance curves for BER and power test (2). Like Figure 8.3(b), the objectives for (a) BER and (b) power are competing. The selection of the population can lead to trends that maximize one at the expense of the other before coming to a compromise.

8.2.4 Throughput

This test analyzes the cognitive engine for optimizing throughput. The test uses the weights from Table 8.7 and produces the results shown in Table 8.8 and in Figure 8.5. These results are straightforward, illustrating the performance of the algorithm to produce a waveform with high data rates and moderately low bit error rate. The balance is achieved through the trade-off provided by using the BER and throughput objectives. We discussed this when defining the throughput objective in Chapter 4.

Table 8.7
Objectives: Throughput

BER	SINR	Thr.	BW	Spec. Eff.	Int.	Pwr.	Comp.
0.75	N/A	0.50	N/A	N/A	N/A	0.50	N/A

8.2.5 Waveform Efficiency

The combination of these objectives leads to the results in Table 8.10 where the waveform provides the desired balance, perhaps with a BER that is slightly

Table 8.8
Waveform Settings and Results: Throughput

Knob	Settings
Modulation	8PSK
Transmit power (dBm)	15.37
Symbol rate (sps)	1.0
Pulse shaping	RRC, 0.56
Normalized frequency	0.229
Packet size	1026

Meters	Sim. Result
BER	2.78×10^{-2}
SINR (dB)	N/A
Spec. Eff. (bps/Hz)	N/A
BW (Hz)	N/A
Throughput (bps)	3.0
Interference (dBm)	N/A
Power (dBm)	15.37
Computation (ticks)	0
Obs. BER	3.965×10^{-3}

Table 8.9
Objectives: Waveform Efficiency

BER	SINR	Thr.	BW	Spec. Eff.	Int.	Pwr.	Comp.
0.90	N/A	0.60	0.50	0.30	N/A	0.40	0.90

larger than it should be. The cognitive engine produced a spectrally small waveform but did not go to an 8PSK modulation to meet the throughput objective, as this would have negatively affected the BER and computational performance. However, the throughput plot from Figure 8.6 shows higher throughput and symbol rates were tried and rejected in the end while the spectral efficiency continuously improved. The power and BER curves show the same trend as in the above example where the middle generations show trends towards lower BER and higher power. The recurrence of this property indicates a performance trend in the algorithm. We previously mentioned concepts of population niching, a concept used to maintain an even distribution of individuals along the Pareto front. The decline of diversity in later generations suggests the need for niching to provide the population diversity that will allow a more complete search and a better set of individuals on the Pareto front from which to select a solution.

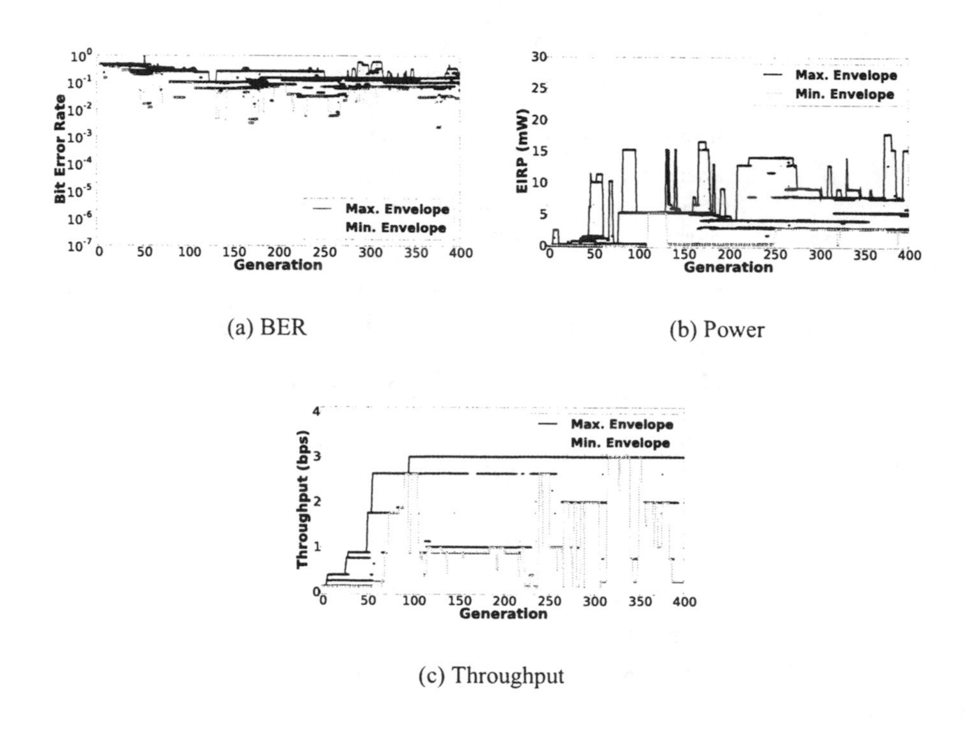

(a) BER

(b) Power

(c) Throughput

Figure 8.5 Performance curves for throughput test. The (c) throughput objective is more directly dependent on the modulation and can increase with less pressure from (a) BER as the (b) power increases.

In none of these problems did the cognitive engine find a waveform that uses differential modulation. This is actually good, because under no circumstances in the system provided does a differential modulation make more sense then a nondifferential modulation. The differential modulations have about a 2-dB loss in the BER performance, and they do not offer any benefit in throughput or bandwidth. Their use in real communication systems is derived from the simpler design requirements for the receiver. However, in the GNU Radio situation, as discussed in Appendix C, the differential modulators only add blocks to the flow graph and so increase the complexity of the system. Therefore, given the current radio platform, there are no situations that will allow a differential modulation an advantage over a nondifferential form, but it is important to point out that this only holds true for the radio platform used. Another platform that implements a simpler differential receiver might show some benefits under certain conditions of low battery capacity.

Table 8.10
Waveform Settings and Results: Waveform Efficiency

Knob	Settings
Modulation	QPSK
Transmit power (dBm)	5.32
Symbol rate (sps)	0.125
Pulse shaping	RRC, 0.10
Normalized frequency	−0.674
Packet size	1405
Meters	**Sim. Result**
BER	2.01×10^{-2}
SINR (dB)	N/A
Spec. eff. (bps/Hz)	3.64
BW (Hz)	0.069
Throughput (bps)	0.25
Interference (dBm)	N/A
Power (dBm)	5.32
Computation (ticks)	5367.67
Obs. BER	2.68×10^{-3}

8.3 Interference Environment

In the next few experiments, we use the full simulation design of Figure 3.5 by introducing the interferers. The radio simulation takes a number to seed the random number generator, thus allowing us to rerun the simulations with the different waveforms. Each simulation uses a random number of interferers from 10 to 15, and the amplitude, frequency, and bandwidth of each interferer is selected at random. These experiments show the ability of the cognitive engine to model waveforms that both meet QoS requirements as well as avoid interferers. We have added a couple of situations that we found illustrated interesting problems in the cognitive engine because of incorrect information received from the sensors.

The experiments are performed by first running the radio simulation with a random seed, collecting the PSD and meters sensor data, reading in the optimization objectives, building a waveform, then rerunning the simulation using the same random seed to test the performance of the waveform under the same simulation conditions.

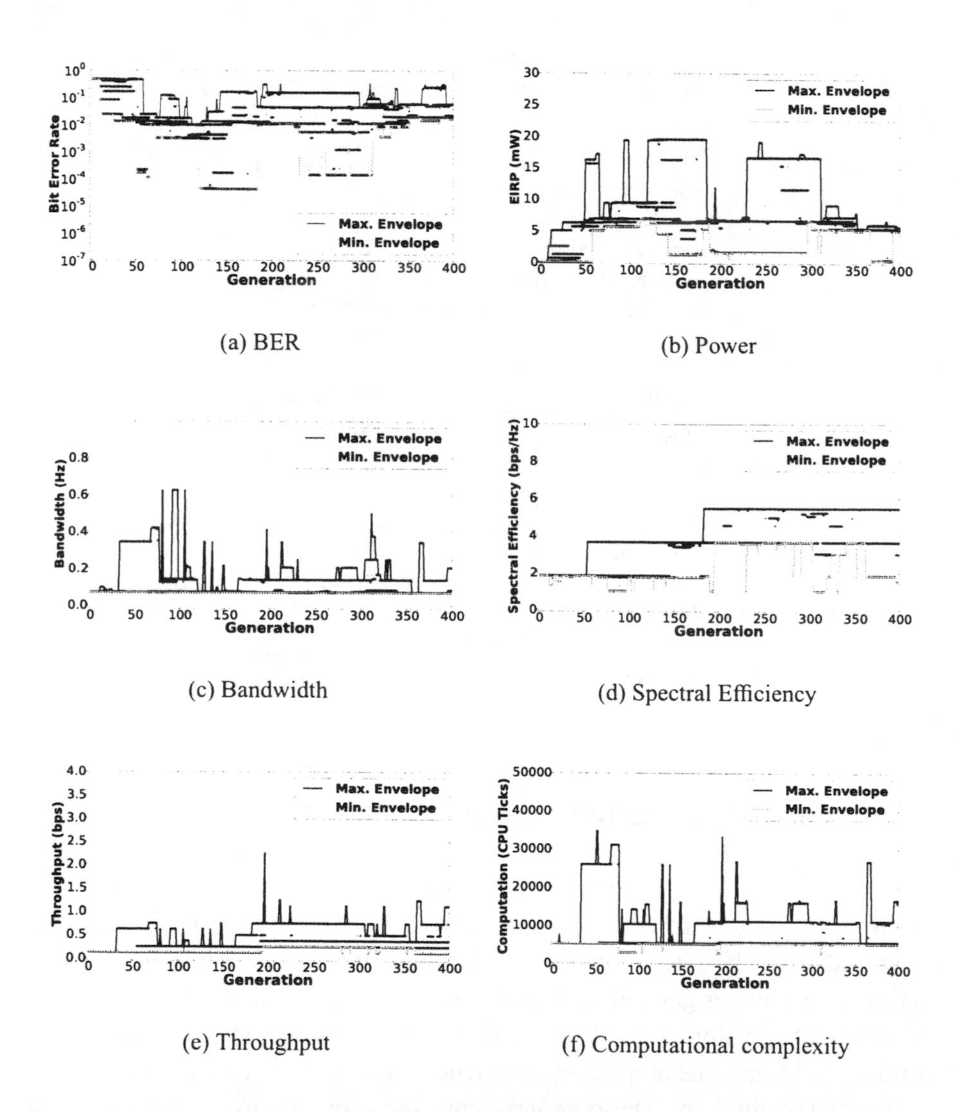

(a) BER

(b) Power

(c) Bandwidth

(d) Spectral Efficiency

(e) Throughput

(f) Computational complexity

Figure 8.6 Performance for waveform efficiency test with plots for objectives (a) BER, (b) power, (c) bandwidth, (d) spectral efficiency, (e) throughput, and (f) computational complexity. The GA searches through a large number of solutions with different objectives dominating at different times before converging to compromised solutions by the final generations.

8.3.1 Interference (1): Simple BER Tests

The first, simplest test under these conditions is to ask the cognitive engine to build a waveform that avoids the interferers and minimizes the BER as represented by the objective weights in Table 8.11. Table 8.12 shows that the

BER objective was met and Figure 8.7 shows the frequency domain of the simulation to demonstrate that the interferers were avoided properly. In the frequency plots, we show the signal seen by the receiver as the combination of the transmitter and interferers and the transmitted signal is shown as the dotted black line that has been adjusted for path loss. Figure 8.8 shows that the objectives in this case were easily met. The SINR increased to its maximum quickly, and the interference power was always very low. This figure indicates that very good solutions were found early and dominated the population. The figure also shows two plots of the interference objective. The plot in Figure 8.8(c) shows the interference using the same axis as all other interference plots while the plot in Figure 8.8(d) shows the same plot zoomed in to better show the objectives performance.

Table 8.11
Objectives: Interference Test (1)

BER	SINR	Thr.	BW	Spec. Eff.	Int.	Pwr.	Comp.
1.00	0.00	0.00	0.00	0.00	1.00	0.00	0.00

Table 8.12
Waveform Settings and Results: Interference Test (1)

Knob	**Settings**
`Modulation`	BPSK
`Transmit power (dBm)`	20.00
`Symbol rate (sps)`	0.750
`Pulse shaping`	RRC, 0.17
`Normalized frequency`	−0.90
`Packet size`	580
Meters	**Sim. Result**
`BER`	1.21×10^{-6}
`SINR (ratio)`	7.12
`Spec. Eff. (bps/Hz)`	1.71
`BW (Hz)`	0.439
`Throughput (bps)`	0.75
`Interference (mW)`	8.63×10^{-5}
`Power (dBm)`	20.00
`Computation (ticks)`	31094.10
Obs. BER	4.5×10^{-6}

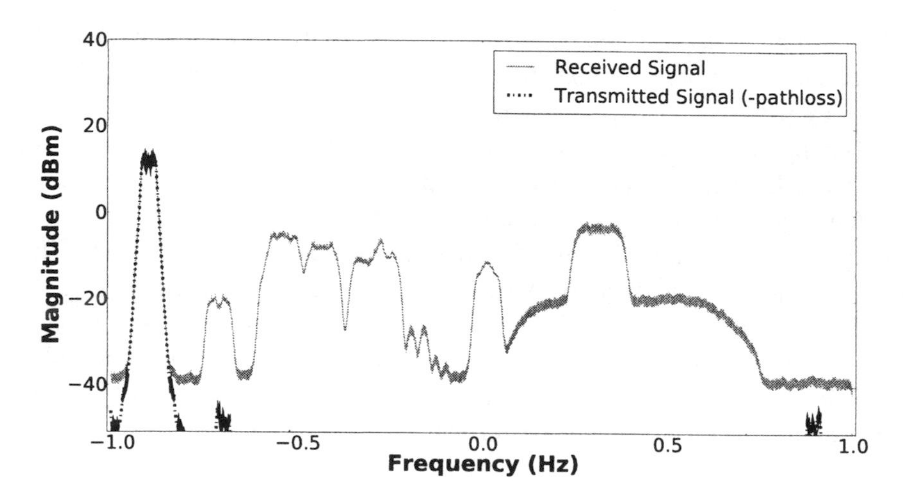

Figure 8.7 Frequency domain plot of interference test (1) where the cognitive engine found a solution that avoids the other users.

8.3.2 Interference (2): Sensor Problems

In the next set of experiments, we use a more interesting set of objectives from Table 8.13 to balance the waveform properties and create waveforms that are bandwidth efficient but have high throughput, low BER, and avoid the interferers. In this first test, the cognitive engine used the information retrieved from the PSD sensor to know where to avoid the interference. As Table 8.14 and Figure 8.9 show, the waveform designed by the cognitive engine had 0 BER and the minimum interference power possible, where 1×10^{-12} is used as the minimum possible value to avoid errors when converting to dBm. However, the observed BER was much higher, and Figure 8.10 shows that the waveform was placed directly on top of one of the interfering signals.

A look at the results returned by the PSD sensor shows why the cognitive engine selected an improper frequency. According to the PSD sensor as seen in Table 8.15, there are only four bands to avoid. The last signal is shown to both start and end where the previous signal ended. In this representation, the interference signal from about 0.55 to 1.0 Hz (normalized frequency) cannot be detected. The cognitive engine, working with this data, did not understand the presence of the signal and therefore could not avoid it. It turns out that the PSD sensor had a small logic error that caused this representation problem, and it was easily corrected. We show this because we want to illustrate this problem to indicate the impact of incorrect information on the cognitive engine; otherwise, under the conditions presented, the waveform optimization looks correct.

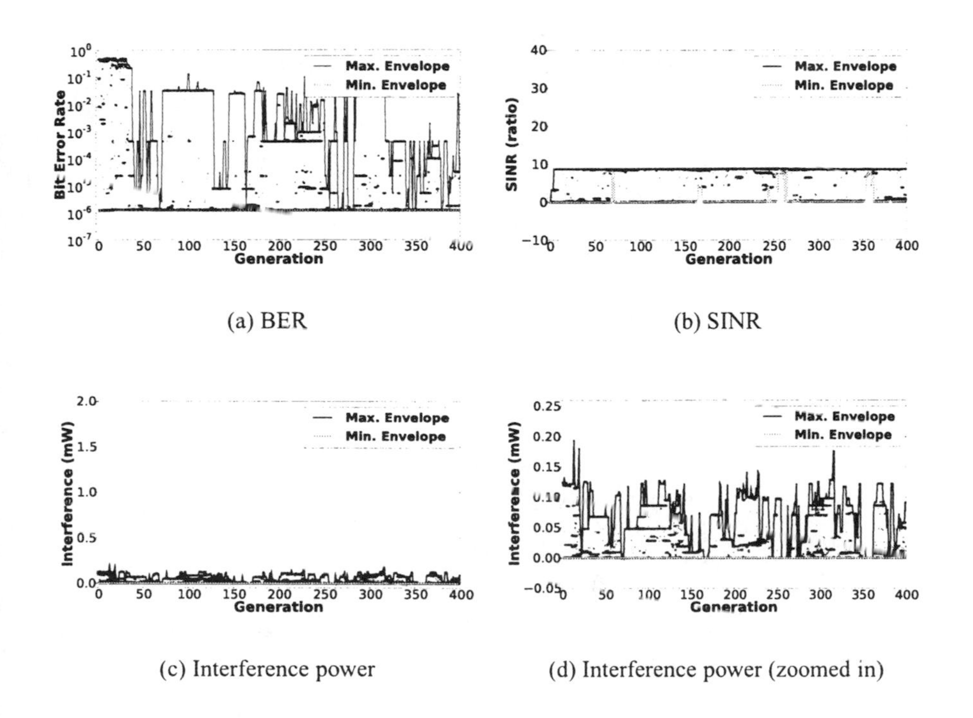

(a) BER

(b) SINR

(c) Interference power

(d) Interference power (zoomed in)

Figure 8.8 Performance curves for interference test (1) showing objective plots for (a) BER, (b) SINR, and (c) interference power. Plot (d) is a zoomed in plot of (c) for a closer look at the variations over the generations.

Table 8.13
Objectives: Interference Test (2)

BER	SINR	Thr.	BW	Spec. Eff.	Int.	Pwr.	Comp.
1.00	0.75	0.60	0.20	0.00	1.00	0.25	0.00

8.3.3 Interference (3): Correcting for Sensors

While the mistake shown in the last example related to a logic error in the sensors, other measurements and uncertainties can also have an effect on the cognitive engine's performance. The meters designed for use with the GNU Radio simulation measure the received power and noise power, and, as Appendix A shows, the measurements are fairly accurate, increasing in uncertainty as the SNR decreases. These tests were conducted outside of the presence of interference sources, which significantly skew the results. The received signal strength of the meters sensor calculates the average magnitude squared of the received signal through the receiver's channel filter.

Table 8.14
Waveform Settings and Results: Interference Test (2)

Knob	Settings
Modulation	8PSK
Transmit power (dBm)	10.02
Symbol rate (sps)	0.750
Pulse shaping	RRC, 0.10
Normalized frequency	0.777
Packet size	383
Meters	**Sim. Result**
BER	0
SINR (ratio)	47109.2
Spec. Eff. (bps/Hz)	5.45
BW (Hz)	0.4125
Throughput (bps)	2.25
Interference (mW)	1×10^{-12}
Power (dBm)	10.02
Computation (ticks)	33430.09
Obs. BER	4.5×10^{-6}

Table 8.15
PSD Sensor Results in Interference Test (2)

Signal	Amplitude (dBm)	f_{min} (Hz)	f_{max} (Hz)
1	−12.12	−1	−0.975
2	−11.45	−0.909	−0.705
3	1.28	−0.332001	0.493
4	−23.02	0.493	0.493

When interference is present, the interference signal is added to the received signal. Because the SNR meter calculation was not designed properly to measure the SINR, the resulting information is significantly skewed when interference is received through the channel filter, specifically by raising the measured signal power and disproportionately affecting the measurements of the path loss and SNR. This information propagates through to the cognitive engine and optimization process. When designing waveforms, the cognitive engine's estimation of the path loss affects the BER measurement, which will then affect what power levels provide acceptable BER values. In this case, because the path loss estimation is significantly decreased due to the increased measurement of the received power, the BER calculation assumes that lower transmit power provides lower BER than it should.

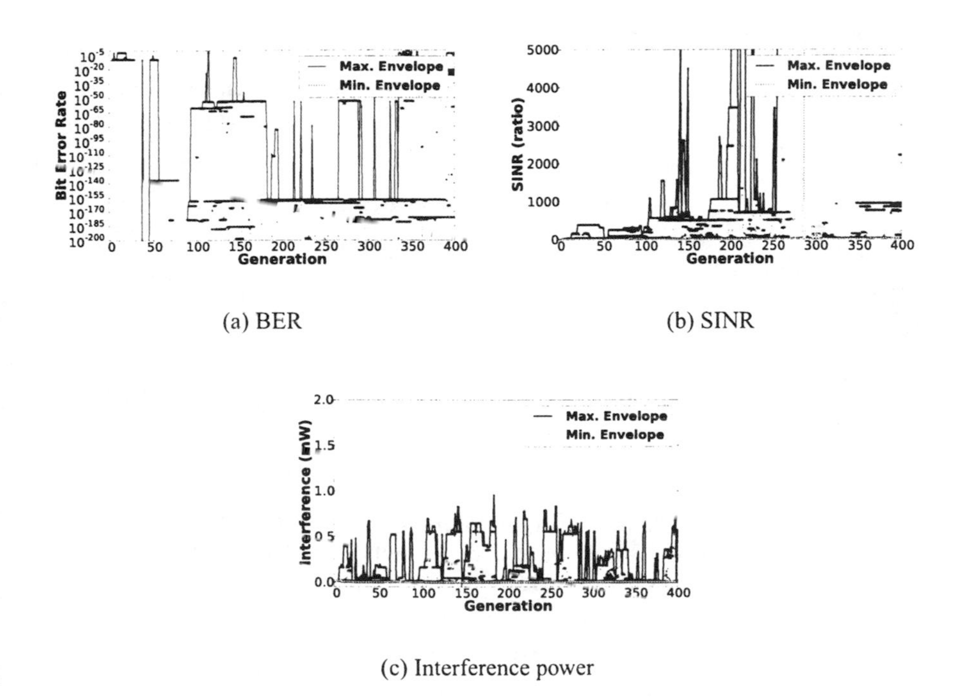

(a) BER

(b) SINR

(c) Interference power

Figure 8.9 Performance curves for interference test (2) showing objective plots for (a) BER, (b) SINR, and (c) interference power. The SINR objective plot (b) shows the mistake where the later generations calculated a very small SINR at the solution chosen.

We used the parameters in Table 8.16 to run the next test. Table 8.17 shows what happens under these conditions. The cognitive engine found that a BPSK signal with a 10.72-dBm transmit power will produce a 0 BER (in other words, a value too small to be represented), which indicates a very large SNR (see Figure A.1 to confirm this). However, with the use of the multiobjective search space, different objectives can pressure the solution into an more acceptable solution. In this case, we use the SINR objective as the pressuring agent in the optimization. When the SINR objective is not used, the power is minimized to 0.1 mW. The results in Figure 8.11 show that using the SINR objective balances the incorrect information that leads to the poor decision based only on the BER. The resulting waveform achieved the desired performance of interference avoidance (see Figure 8.12), low BER, and mid-ranged bandwidth and power.

The problem that the cognitive engine faces is misinformation from the sensor that measures the received signal strength and noise power. The

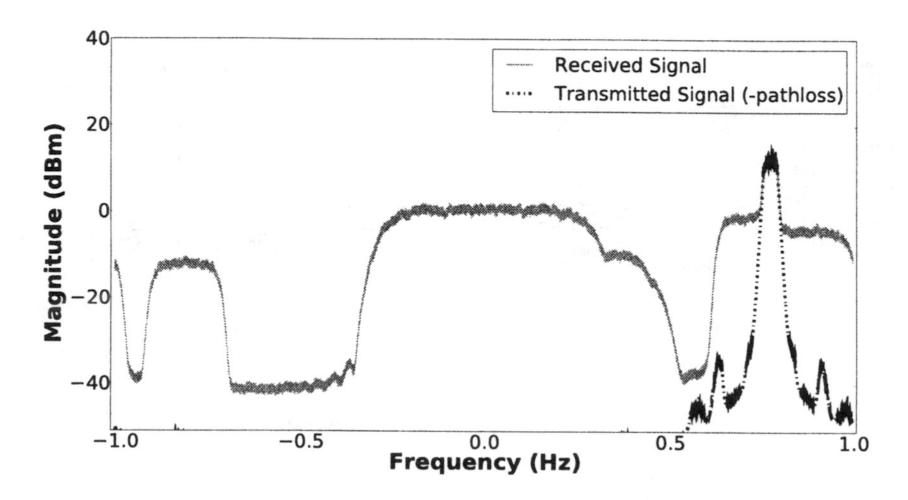

Figure 8.10 Frequency domain plot of interference test (2). A mistake in the interference sensor lead the algorithm to incorrectly believe this spectrum was unoccupied.

Table 8.16
Objectives: Interference Test (3)

BER	SINR	Thr.	BW	Spec. Eff.	Int.	Pwr.	Comp.
1.00	0.75	0.00	0.20	0.00	1.00	0.25	0.00

underlying calculations were set up to measure these values accurately without the presence of an interferer, but when an interferer is present in the received channel, the signal power calculation incorrectly computes the received signal power as much higher than it really is. While we find the behavior and capabilities of the cognitive engine interesting and a valuable experience to enhance the overall understanding of the radio, we want to address how this error can be corrected. The other examples and results in this chapter show that, given proper information about signal and noise power, the cognitive engine properly optimizes and finds waveforms that fit the objectives. Therefore, the cognitive engine needs a sensor capable of properly measuring the signal to interference ratio (SIR) or SINR.

One method of calculating the SINR is through the use of known symbols such as a sequence of training symbols or a known preamble. The GNU Radio implementation uses a known access code, or unique word, that the nodes correlate against to know the start of a packet. This known access code can also provide the required information to calculate SINR. The SINR meter probe is given the modulated access code and correlates against

Table 8.17
Waveform Settings and Results: Interference Test (3)

Knob	Settings
Modulation	BPSK
Transmit power (dBm)	10.72
Symbol rate (sps)	1.000
Pulse shaping	RRC, 0.10
Normalized frequency	−0.90
Packet size	855
Meters	**Sim. Result**
BER	0
SINR (ratio)	22.49
Spec. Eff. (bps/Hz)	1.82
BW (Hz)	0.4125
Throughput (bps)	0.75
Interference (mW)	1.16×10^{-2}
Power (dBm)	10.72
Computation (ticks)	31094.10
Obs. BER	4.61×10^{-6}

the received time domain signal from the channel filter. The output of the correlation spikes when the transmitted access code is received. The known access code sequence can then be subtracted from the received signal to leave any interference signals and the AWGN noise. Before subtracting the known access code, an autocorrelation is performed on it to determine its maximum correlation value. The ratio of the autocorrelation value to the peak of the cross-correlation gives the pathloss assuming the pathloss is constant during the transmission of the access code. This ratio is used to adjust the amplitude of the known sequence to properly remove it from the received signal. Taking the average magnitude squared of this signal yields the interference pulse noise power. The remaining signal after subtraction can then be subtracted from the original received signal to leave only the transmitted signal. The average magnitude squared of this signal is the received signal strength. This ratio of these two power values is the SINR.

Appendix F provides more detail and the mathematical explanation of this meter probe as well as simulated results that show the sensor's proper behavior. This sensor can then be used as a means of calculating the received signal power, the interference plus noise power (or just noise power if no interference is present), and the pathloss, thereby replacing the signal power and noise power probes used previously in this work.

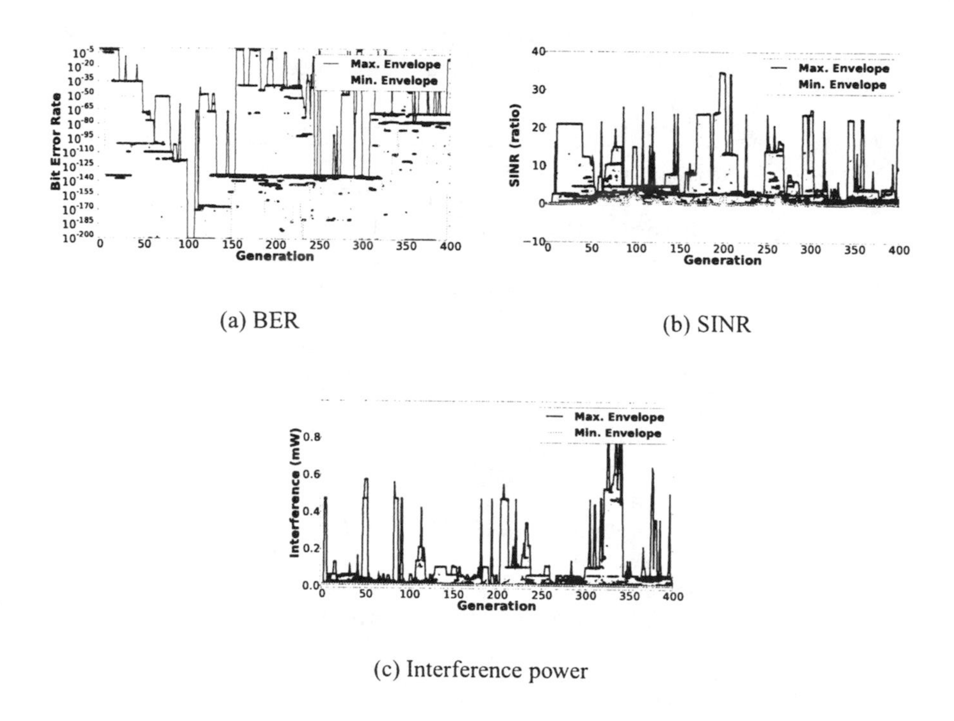

(a) BER

(b) SINR

(c) Interference power

Figure 8.11 Performance curves for interference tests (3) showing objective plots for (a) BER, (b) SINR, and (c) interference power. The SINR calculations are unusually high due to the miscalculation of interference power in the signal power.

8.3.4 Interference (4): Throughput with Low Spectral Footprint

We revisit the second interference experiment that had the goal of providing a high throughput, low BER, and low spectral footprint while avoiding the interference but had trouble because of the bad sensor. This experiment uses the parameters in Table 8.18 and the information provided by the problems of the last two tests to see how well the cognitive engine really performs for this task. The resulting waveform attained all of the desired performance, as shown in Table 8.19, Figure 8.13, and Figure 8.14, with a moderate throughput and small bandwidth by using a high-order modulation (8PSK) with a small bandwidth while avoiding the interferers.

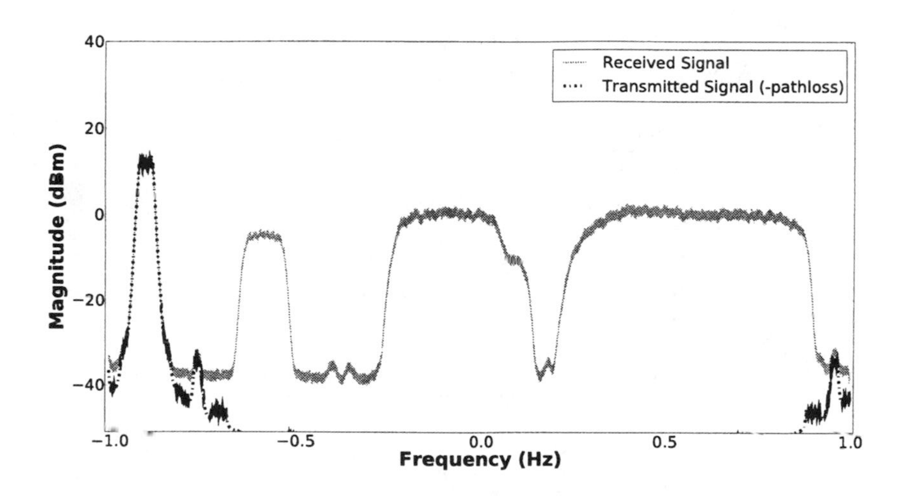

Figure 8.12 Frequency domain plot of interference test (3).

Table 8.18
Objectives: Interference Test (4)

BER	SINR	Thr.	BW	Spec. Eff.	Int.	Pwr.	Comp.
1.00	0.75	0.40	0.20	0.00	1.00	0.25	0.00

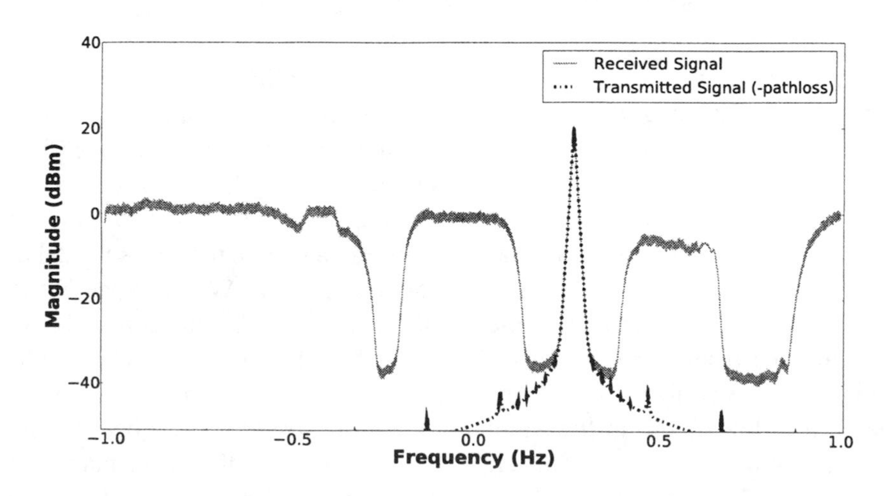

Figure 8.13 Frequency domain plot of interference test (4). With the correctly operating sensor, the cognitive engine finds open spectrum in which to transmit.

Table 8.19
Waveform Settings and Results: Interference Test (4)

Knob	Settings
Modulation	8PSK
Transmit power (dBm)	16.73
Symbol rate (sps)	0.125
Pulse shaping	RRC, 0.10
Normalized frequency	0.275
Packet size	1472
Meters	**Sim. Result**
BER	0
SINR (ratio)	289153
Spec. Eff. (bps/Hz)	5.45
BW (Hz)	0.0688
Throughput (bps)	0.375
Interference (mW)	1.0×10^{-12}
Power (dBm)	16.73
Computation (ticks)	5571.81
Obs. BER	0

8.4 Case-Based Decision Theory Example

The case-based system has large potential for learning and improving the performance of the cognitive engine. The case-based system can feed not only previous solutions, but also adjustments to the optimization parameters like the GA population size and terminating conditions. Our goal here is to provide the system for this kind of research to continue. To just get a feel for how the case-based learning system can be applied, we ran a simple test. This test takes the problem of the second BER and power optimization. We use a weight of 0.60 for the BER objective this time to balance the two objectives for a better solution. The results are shown in Figure 8.15 and Table 8.20. The results show that the waveform is well representative of the problem specifications. The BER is low (and 0 in the simulation) and the power was higher, but not at its maximum. Figure 8.15 provides a better clue into the performance of the case-based system. Unlike the performance plots of Figure 8.4, the initial population shows better distribution in the search space with initial members closer to the final results. This population could have easily been terminated much earlier and achieved a good solution.

Although not a rigorous test of the case-based system, the initial system works and shows promise for performing learning and future potential. The

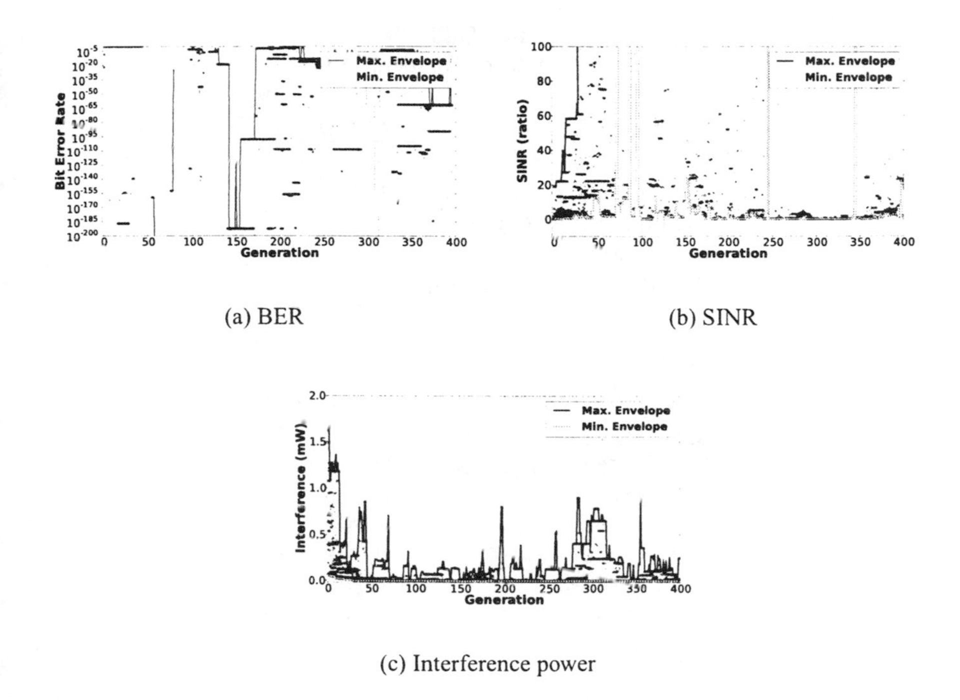

(a) BER

(b) SINR

(c) Interference power

Figure 8.14 Performance curves for interference tests (4) showing objective plots for (a) BER, (b) SINR, and (c) interference power. The (b) SINR and (c) interference curves show good convergence at low values as they are working with correct information.

implementation in the cognitive engine will allow for more advanced concepts and implementations as the research develops.

8.5 Over-the-Air Results

The first over-the-air tests were performed during the IEEE Conference on Dynamic Spectrum Access Networks (DySPAN) in April of 2007 in Dublin, Ireland [1]. At the conference, a number of participating companies and research labs set up and ran their cognitive radio and dynamic spectrum access equipment using spectrum specifically licensed by Ireland's Communications Regulators (COMREG). We set up two of our cognitive radio nodes at the conference using the GNU Radio SDR platform with USRP RF front ends, a PSD sensor, and the WSGA optimizer. The implementation of this cognitive engine is shown in Figure 8.16, which uses a sensor to pull in the PSD

Table 8.20
Waveform Settings and Results: CBDT-GA

Knob	Settings
Modulation	BPSK
Transmit power (dBm)	15.93
Symbol rate (sps)	0.250
Pulse shaping	RRC, 0.69
Normalized frequency	−0.017
Packet size	275
Meters	**Sim. Result**
BER	1.51×10^{-4}
SINR (ratio)	N/A
Spec. Eff. (bps/Hz)	N/A
BW (Hz)	N/A
Throughput (bps)	N/A
Interference (mW)	N/A
Power (dBm)	15.93
Computation (ticks)	N/A
Obs. BER	0

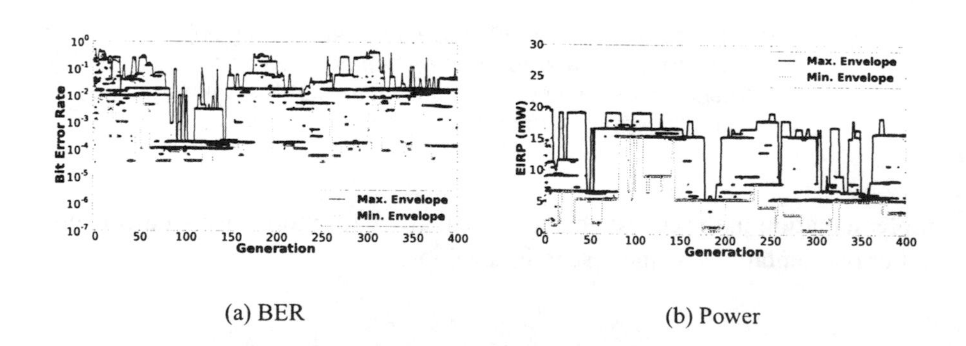

(a) BER (b) Power

Figure 8.15 Performance curves for the case-based application to the second BER and power test. The (a) BER and (b) power objectives were obtained about 50 generations into the optimization process. The remaining generations just balanced the objectives back and forth and provide no new benefit.

information, the WSGA, and the verification system to ensure regulatory compliance of the waveforms.

The spectrum regulations are shown in Table 8.21, although we limited the cognitive radio to use the bands centered at 406.9750 and 408.7750 MHz

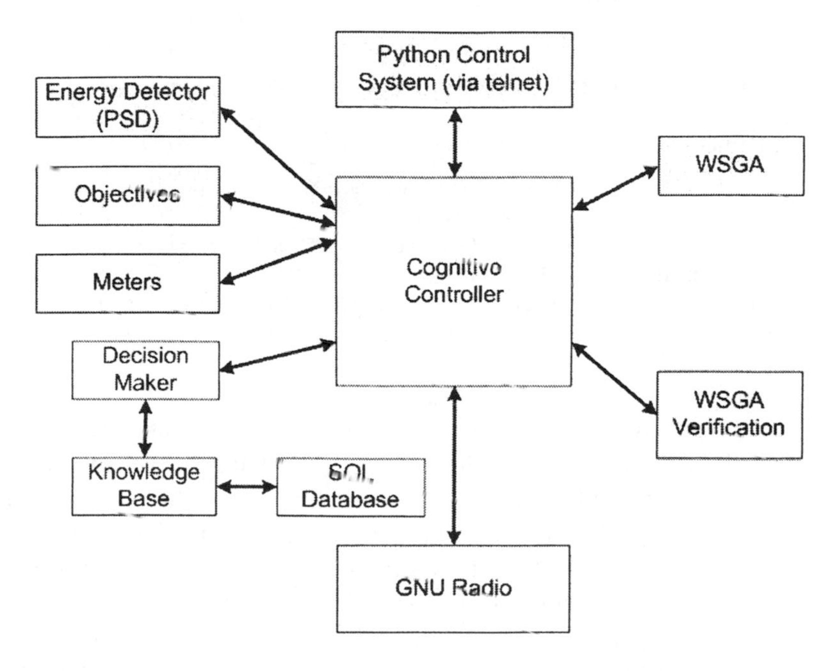

Figure 8.16 The cognitive engine design for online implementation.

to limit the possible range and for a more interesting environment.

Table 8.21
Frequency Allocations at IEEE DySPAN, 2007

Channel	Center Freq. (MHz)	Max ERP	BW (MHz)
1	231.2250	1 W (0 dBW)	1.75
2	233.0250	1 W (0 dBW)	1.75
3	234.8250	1 W (0 dBW)	1.75
4	236.6250	1 W (0 dBW)	1.75
5	238.4250	1 W (0 dBW)	1.75
6	386.8750	1 W (0 dBW)	1.75
7	396.8750	10 W (10 dBW)	1.75
8	406.9750	1 W (0 dBW)	1.75
9	408.7750	10 W (10 dBW)	1.75
10	436.8750	1 W (0 dBW)	1.75
11	2056.000	1 W (0 dBW)	50.0
12	2231.000	1 W (0 dBW)	50.0

Table 8.22 shows the knobs available for the over-the-air experiments.

During the demonstration, of the two cognitive radios, one was the master that would design a waveform and push it to the other. The master

Table 8.22
Knobs Available to the GNU Radio: Over-the-Air Experiments (1)

Knob Name	Knob Settings
Modulation	DBPSK, DQPSK, GMSK
Transmit power	0 - 20 (dBm)
Symbol rate	125 - 500, steps of 25 (kbps)
Pulse shaping	0.1 - 1.0, steps of 0.01
Center frequency	406.1×10^6 - 409.65×10^6, steps of 1 (kHz)
Frame size	100 - 1500, steps of 1

radio would first sweep the spectrum, determine if any other radios were present, and then design a waveform to fit the channel. We attempted to perform a streaming audio service across the two nodes that would provide a high throughput, low error connection amidst the other radios operating in the same frequency. Figure 8.17 shows the results of one of the sensing and optimization processes that occurred during the conference.

In this figure, the spectrum regulations are shown as the thick black masks that surround the available spectrum in frequency and power. The figure shows that during one of the spectrum sweeps an interfering node was present in the middle of the allocated spectrum. The resulting waveform found a position in the spectrum that was allowed in both frequency and power and did not overlap the interfering signal. Furthermore, this signal was a 250-kbps QPSK waveform that provided adequate quality of service for the audio application.

To test the latest version of the cognitive engine, we use operating frequencies between 2.405 to 2.415 GHz with three interfering radios. Two of the interference nodes are 1-MHz wide QPSK signals generated from the Centre for Telecommunications Value-Chain Research (CTVR) IRIS software radio [2] and a third is a 1-MHz wide OFDM signal generated using the Anritsu MG3700A signal generator. The signals were positioned at 2.4075 GHz (IRIS QPSK 1), 2.410 GHz (IRIS QPSK 2), and 2.4125 GHz (signal generator OFDM). The cognitive radio node has the waveform capabilities described in Table 8.23.

When asked to design a waveform using the same objectives as Table 8.18, the cognitive engine produced a 200-kbps QPSK signal with a 12-dBm transmit power. At this point, the performance of the cognitive engine is well understood in building signals that can produce high data rates with low BER. The most interesting performance property of this particular example is that the PSD sensor accurately modeled the interference environment, and the cognitive radio optimized around these interferers. Figure 8.18 shows

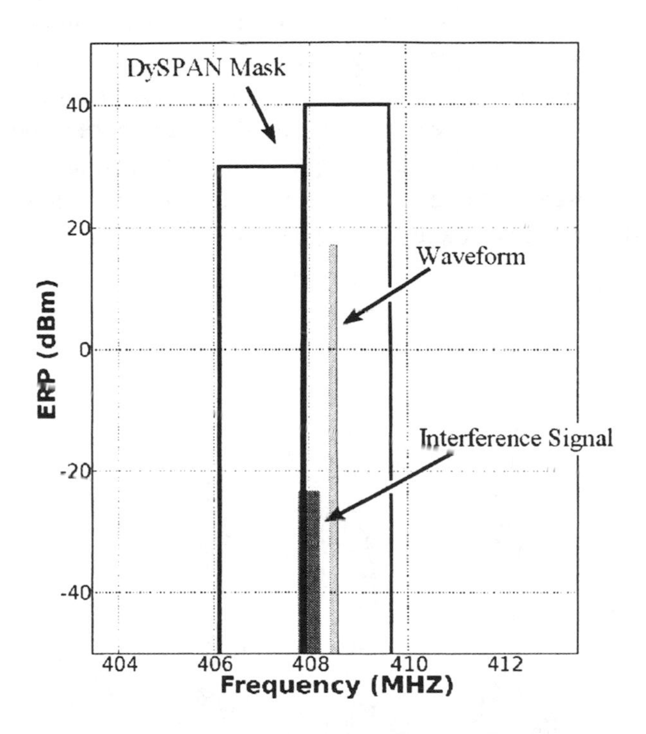

Figure 8.17 DySPAN 2007 spectrum measurements of an interferer and the cognitive radio transmitting in a given spectrum mask.

Table 8.23
Knobs Available to the GNU Radio: Over-the-Air Experiment (2)

Knob Name	Knob Settings
Modulation	DBPSK, DQPSK, GMSK
Transmit power	0 - 20 (dBm)
Symbol rate	15 - 500, steps of 10 (kbps)
Pulse shaping	0.1 - 1.0, steps of 0.01
Center frequency	2405×10^6 - 2415×10^6, steps of 1 (kHz)
Frame size	100 - 1500, steps of 1

the spectrum as captured by an Anritsu Signature signal analyzer. The three interfering signals are seen at 2,4075, 2,410, and 2,4125 MHz and the cognitive engine's waveform is located to the left of all three interferers at 2,4057 MHz. The signal on the right edge of the plot around 2,414 MHz was not part of the experiment but a random signal picked up in the unlicensed band while collecting the data. Figure 8.19 shows the optimization curve for the interference objective. In the early generations, the interference power for many of the solutions is very large, but the heavy selection pressure to minimize the interference allows the cognitive engine to quickly find spectrum free of interference. Because there was a large amount of open space in the 10 MHz of spectrum used in this example, it was fairly easy for the GA to converge on a good solution within about 50 generations.

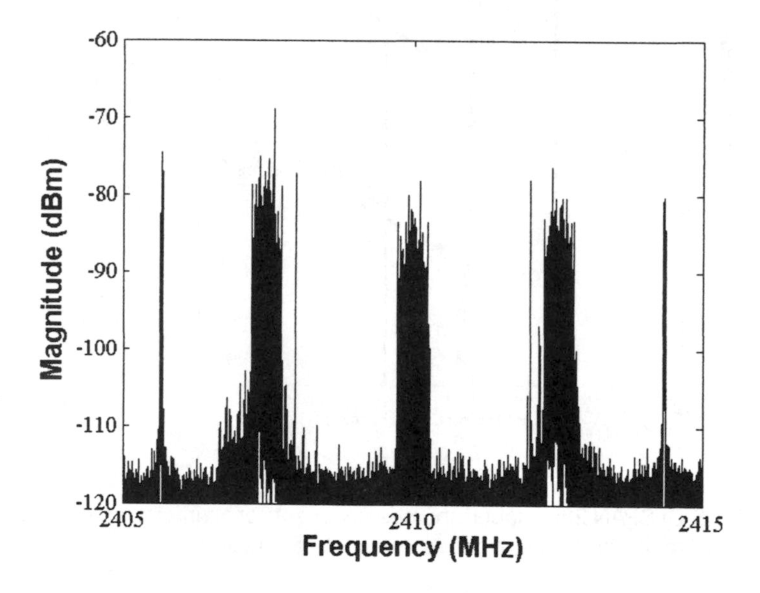

Figure 8.18 Frequency domain plot of over-the-air test (2). The cognitive radio is transmitting at about 2,406 MHz with three known high-bandwidth interferers and a fourth unknown interferer at around 2,414 MHz.

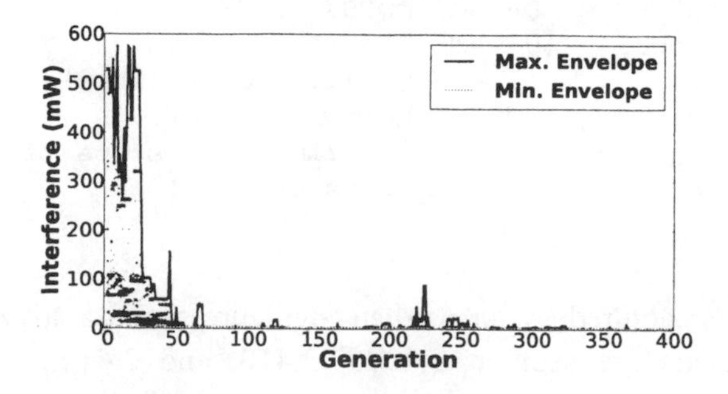

Figure 8.19 Interference performance curves for the over-the-air experiment (2).

Unfortunately, it is difficult to capture the real performance of the system online except for providing example cases. The important lesson of this experience is the ability of the cognitive engine to move from the simulation environment that can be used to more adequately understand the behavior and operation of the engine to the online system using real radio hardware. The flexible and extensible cognitive engine has been shown through these experiments.

8.6 Conclusions

We demonstrated the use of the cognitive engine operating in both a simulation and a real-world demonstration. The modular component structure of the cognitive engine enables designers to build new components for different purposes. In the simulation environment, a simple component allows the interface between the cognitive engine and the radio simulation platform. It also connects the PSD and meters sensors that communicate with the simulation to collect the data and pass it to the cognitive engine. In the online version, the components were replaced with others that enable communications with real radio systems. As we discussed in this chapter, incorrect information can have a significant effect on the cognitive engine's performance, such as the logic error in the PSD sensor or the miscalculation of the SNR in the presence of interferers with the meters sensor. The component structure makes unit testing of each of these pieces simple in order to work out bugs, and components are easily replaced when better versions are made available.

Although it is difficult to show the performance of the cognitive engine, we worked through some simple examples to understand the general trends of the optimization process. Early simulations showed how the cognitive engine optimizes for such objectives as BER and power performance. Later examples that use many more objectives make the analysis much more difficult in determining the performance by looking at each individual objective. In these cases, the multiobjective Pareto front surface is the important measure, but difficult to illustrate. Instead, we showed some performance results and waveforms produced under certain objectives. The results first showed that the cognitive engine can successfully optimize waveforms given performance requirements. Also, many lessons can be learned from these results and the performance plots. We discussed some of the issues regarding poor information received from the sensors and pointed out where some advanced topics in the genetic algorithm design and optimization would be useful.

The system demonstrated in this chapter is the culmination of the theory and discussions of the preceding chapters. While much of the theory comes from many different disciplines and applications, the intent of this book was to discuss the application of these concepts to wireless communications systems. This chapter provides that necessary bridge to the real implementation of a cognitive radio as well as demonstrations of it working. The next chapter wraps up and looks at how this work can be extended in the future.

References

[1] D. C. Sicker, "The Technology of Dynamic Spectrum Access and Its Challenges," *IEEE Communications Magazine*, Vol. 45, No. 6, pp. 48 – 48, 2007.

[2] P. Sutton, L. Doyle, K. E. Nolan, "A Reconfigurable Platform for Cognitive Networks," *IEEE Proc. Cognitive Radio Oriented Wireless Networks and Communications (CROWNCOM)*, Jun. 2006, pp. 1 – 5.

9

Conclusions

In this book, we have discussed and presented different methods for how artificial intelligence can improve communications. We applied these concepts to build the enabling technology of cognitive radio, the cognitive engine. Throughout, we explained what cognitive radio and a cognitive engine are, as well as the software radio platform used to realize the actions of the cognitive engine. The most significant developments are the discussion and theoretical analysis in Chapter 4 of using multiobjective optimization to perform the cognitive actions. Chapter 5 showed an algorithm suited to perform the multiobjective analysis and optimization, followed up by an improved learning system using case-based decision theory in Chapter 6. We presented a brief but important treatment of using the cognitive engine as part of a network of cognitive radios and some considerations for what information can be distributed to all nodes on a network. We then showed the implementation of the cognitive engine to adapt the GNU Radio system under certain system conditions.

The intent of this book was to bring together the theory of the cognitive engine with a working model suitable for operation. The development of the multiobjective optimization was necessarily limited to the SDR platform available, but through this, we laid down the fundamentals of developing and analyzing a cognitive engine. Through the development of the distributed cognitive engine, we provide a system that can be extended, enhanced, and made usable for future applications and radio systems. One of the most significant challenges of our work was building a system that both shows the operation of the theory but is also usable and reusable in further developments by using generic, documented interfaces and simple script languages to enable ease of use.

Because the intentions of our work were to provide a usable system on the SDR technology we had available, there are many advances not considered in the development and analysis. There are a few significant areas that we left out of the analysis that we wish to address briefly as part of future research. First, we will discuss the application of the cognitive engine to optimize multicarrier systems. Second, the significant work done in adaptive systems has thus far been ignored, but we will discuss how this work can be used with and by a cognitive radio system. Finally, we address some of the AI and learning issues that have been brought up throughout this document that can enhance the future capabilities of the cognitive engine.

9.1 Application to Multicarrier Waveforms

One of the more popular subjects in communications, for physical, MAC, and network layer researchers, is multicarrier systems. In particular, orthogonal frequency division multiplexing (OFDM) dominates this discussion, and we will use it here as an example, although much of this discussion should be easily extensible to other multicarrier techniques. In OFDM, each symbol is carried over a number of orthogonal subcarriers. Subcarriers can be used in different ways. Some subcarriers can be used for pilot tones to help with equalization at the receiver, while other subcarriers can each transmit information using a different modulation. Subcarriers can be used or unused to help shape the spectrum, and each subcarrier's bandwidth can be altered to provide different spectral properties and communications capabilities. As an example, the IEEE 802.16 standard uses OFDM and provides a wide range of adaptable parameters [1].

OFDM modulation has great potential for cognitive radio systems through all of the adaptable parameters that the radio can use to change the behavior and performance. The length of the cyclic prefix balances the spectral efficiency with the multipath resistance of the signal, and the bandwidth of each subcarrier changes the data rate while altering the protection against frequency-selective fading. Equation (9.1) shows how parameters can be adjusted to affect the data rate of an OFDM waveform where B is the symbol bandwidth, L is the number of subcarriers, L_d is the number of data subcarriers, M is the modulation order, and G is the guard fraction (ratio of the cyclic prefix length to the total symbol length) [1]. Different modulations can be used per subcarrier to balance the properties of the modulation against the bit error rate. Channel coding, too, provides a way of balancing error correcting capabilities with data rate. The setting of each of these parameters depends on the channel conditions in the same way as in the narrowband signal analysis we provided earlier. The analysis and implementation of the cognitive engine, however, is easily enhanced to use multicarrier techniques by adding the sensors required to understand the channel conditions (e.g., multipath and fading) and the objective functions to properly model the effects of the parameters.

$$R = \frac{B}{L} \frac{L_d \log_2(M)}{1 + G} \quad \text{bps} \tag{9.1}$$

9.2 Strategies, Not Waveforms

The optimization process we developed focused on finding values for the knobs in order to satisfy certain quality of service objectives. However, the cognitive engine reacts to changes in the environment and user needs, so the adaptation is situational and does not act on a packet-by-packet time scale. On the other hand, researchers in communications have developed sophisticated and powerful techniques for locally-adaptive schemes; that is, methods that focus on adaptation of certain properties to enhance communication performance. Such techniques include adaptive power control, as in traditional cellular telephony, or adaptive modulation as seen in many standards such as Universal Mobile Telecommunications System (UMTS), IEEE 802.11, and IEEE 802.16. The 802.16, or WiMAX, standard has many interesting possibilities in this area because of the significant number of adaptive parameters available.

The future cognitive radio should take advantage of all of these techniques to build the communications system. Instead of trying to find the best power to the tenth of a dB, the cognitive engine could instead choose an adaptive power strategy that makes sense for the current conditions. Likewise, the same can be done for modulation, channel code adaptations, or dynamic spectrum access technologies. We like to think of the cognitive radio as developing a strategy to work within an environment and for a particular service. Instead of designing a waveform, the cognitive engine builds a strategy.

The objective functions we presented in Chapter 4 were based on the concept of waveform adaptation, and we showed how each interacts with the entire system. Building a cognitive engine to work with adaptive strategies is even more complicated since each adaptive technique would interact with the overall quality of service of the system. We argue that the premise of the cognitive engine stays the same as we have developed it here. Instead of genes representing particular parts of a waveform and objective functions analyzing each waveform, the genes represent adaptive strategies, and so the objective functions would need to be developed to properly analyze and model the behavior of the strategy. Furthermore, this type of adaptation places more responsibilities on the radio platform to support many of these possible techniques.

9.3 Enhanced Learning Systems

When discussing the individual topics throughout this document, we have pointed out advances to the theory that others have investigated, such as advanced techniques for genetic algorithms. We would like to address a few of these here, now that we have described the entire system. The areas of particular interest are potential enhancements to the multiobjective optimization, the genetic algorithm, and the case-based learning system.

The multiobjective optimization work has provided an analysis of the individual objectives used as well as different methods of aggregating the objectives to allow comparisons. In this discussion, we briefly pointed to the lessons learned from the work of economists in modeling and analyzing utility and production functions. While there are many different methods of comparing solutions, we only really investigated a couple of them. There is still a significant amount to be learned from the economics literature, and much analysis left to understand how to best apply these concepts to the cognitive engine.

We have tried to avoid including too much domain knowledge when developing and analyzing the performance; instead, we prefer to make direct comparisons of performance in each objective. While our attempts have been for the purpose of increasing generality of the system's operation, specific analysis and domain knowledge could enhance solutions. We pointed to some work done on fuzzy logic cognitive radios [2] as this might be a technique to allow a trade-off between generality and domain-specific solutions. The fuzzy system can be programmed to include basic communication system rules to guide the development of solutions and offer bounds and aggregation techniques to the objective analysis. Fuzzy rule sets could establish basic performance metrics, such as maximum BER values acceptable for certain optimization goals. Using fuzzy logic could enable an aggregation function to combine objectives into a single metric for performance comparison.

The genetic algorithm literature analyzes many advanced topics [3, 4, 5, 6]. In particular, we pointed out the concept of parallel genetic algorithm analysis in Chapter 7, but any conference proceedings, book, or journal on genetic algorithms lists many other advanced techniques. Adaptive parameters are a popular method of improving performance by changing the crossover or mutation rates, depending on the trends of the algorithm performance. We have presented one method of seeding the population using a case base of past solutions. In this discussion of Chapter 6, we also alluded to many other advances the case-based feedback mechanism could offer, including such parameter adjustments like population size, mutation rates, and termination conditions. There are many gains available in the optimization process, and the techniques mentioned here are just some of the low-hanging fruit available.

The case-based system offers other performance improvements beyond its control of the genetic algorithm. We presented many of these in Chapter 6, but we will reintroduce a few here. The case-based decision theory mechanism depends greatly on the similarity, utility, and decision-making functions, as we showed during the results of the knapsack problem. This warrants further study into these for application to the cognitive engine. There are also many tunable parameters to study in the case-based design, including the size of the case base, the number of solutions to pull from the case base, and from where to find these solutions. Another aspect that we have discussed is the method of remembering and forgetting solutions in the case base, as well as potentially using multiple types of case bases to realize concepts like short-term and long-term memory. There is still a lot research to be done in these applications.

9.4 Final Thoughts

In "Computing Machinery and Intelligence," Alan Turing built the foundations of artificial intelligence [7]. In it, he concludes, "we can only see a short distance ahead, but we can see plenty there that needs to be done," a conclusion that any good researcher will reach. While there may be little ground left to be covered in individual physical layer concepts such as modulation and coding, we have not yet fully tapped the potential of the communications system as a whole. The interactions among all aspects of a waveform and communications system, as well as the behavior of networks, are rich fields of research that we are only now developing. Our goal here was to present the methods by which a radio can intelligently analyze and build systems for its own purpose. As we have pointed out in this chapter, there remains much to be done to enhance our current communications platforms to provide easy, ubiquitous communications and access to information.

References

[1] J. G. Andrews, A. Ghosh, and R. Muhamed, *Fundamentals of WiMAX: Understanding Broadband Wireless Networking*, Upper Saddle River, NJ: Prentice Hall, 2007.

[2] N. Baldo and M. Zorzi, "Fuzzy Logic for Cross-Layer Optimization in Cognitive Radio Networks," *IEEE CCNC*, Jan. 2007, pp. 1128 – 1133.

[3] D. E. Goldberg, *Genetic Algorithms in Search, Optimization, and Machine Learning*, Reading, MA: Addison-Wesley, 1989.

[4] L. Chambers, *Practical Handbook of Genetic Algorithms: New Frontiers*, Boca Raton, FL: CRC Press, 1995.

[5] M. Srinivas and L. M. Patnaik, "Adaptive Probabilities of Crossover and Mutation in Genetic Algorithms," *IEEE Transactions on Systems, Man and Cybernetics*, Vol. 24, pp. 656 – 666, 1994.

[6] Y. J. Cao and Q. H. Wu, "Convergence Analysis of Adaptive Genetic Algorithms," *IEE Proc. Genetic Algorithms in Engineering Systems: Innovations and Applications*, 1997, pp. 85 – 89.

[7] A. M. Turing, "Computing Machinery and Intelligence," *Mind*, Vol. 59, pp. 433 – 460, 1950.

Analysis of GNU Radio Simulation

Chapter 3 introduced the GNU Radio and the modulation schemes available. The chapter also presented the simulation environment and the method used to collect the performance meters, which are required by the cognitive engine This appendix provides a performance analysis of the simulation environment we developed by plotting the BER versus E_b/N_0 curves for the supported modulations of GMSK, BPSK, QPSK, 8PSK, DBPSK, DQPSK, and D8PSK. The plots are developed by measuring the simulation performance of the SDR code used by the cognitive engine as well as the methods used to set the signal power, propagation and channel conditions (noise power and path loss), and signal, noise, and BER estimations at the receiver.

A.1 Bit Error Rate Plots

The first order of analysis is to see how well the modulators and demodulators work under known conditions; in particular, we are looking at how closely these components match the the theoretical performance in AWGN channels. Equations (4.8) through (4.12) provide the theoretical BER equations. Figures A.1 through A.7 show the simulated BER compared to the theoretical curves. These figures also provide an analysis of how well the E_b/N_0 calculations perform by plotting the known E_b/N_0 (by setting the noise level and path loss) against the estimated E_b/N_0. The error bars in these figures represent one standard deviation from the estimated mean over 20 trials.

Unfortunately, there is no available theoretical equation of D8PSK, so as a crude approximated lower bound, we have used the theoretical curve for 8PSK and assumed that the D8PSK curve would have approximately a 2-dB worse performance like other differentially coded modulations.

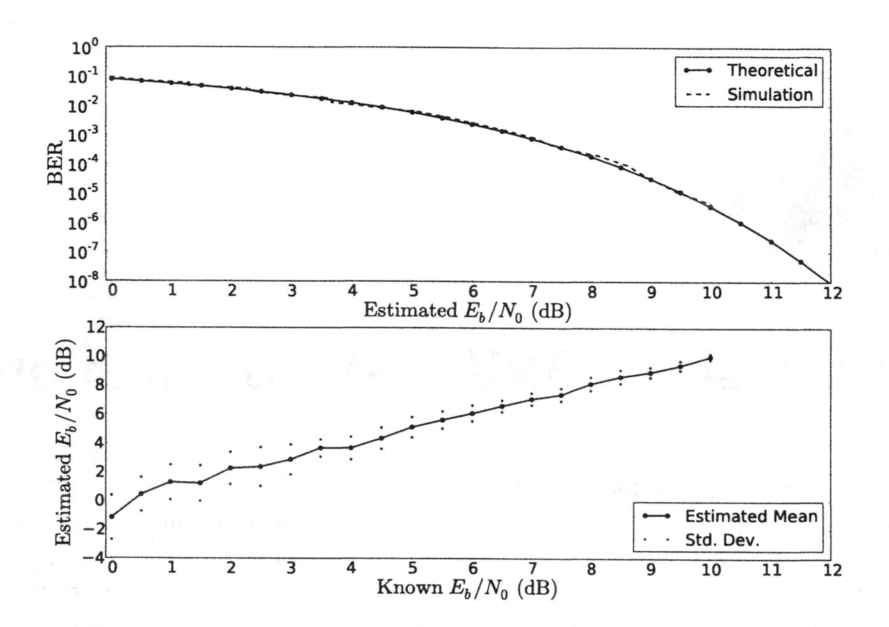

Figure A.1 BER curves and E_b/N_0 plots for BPSK.

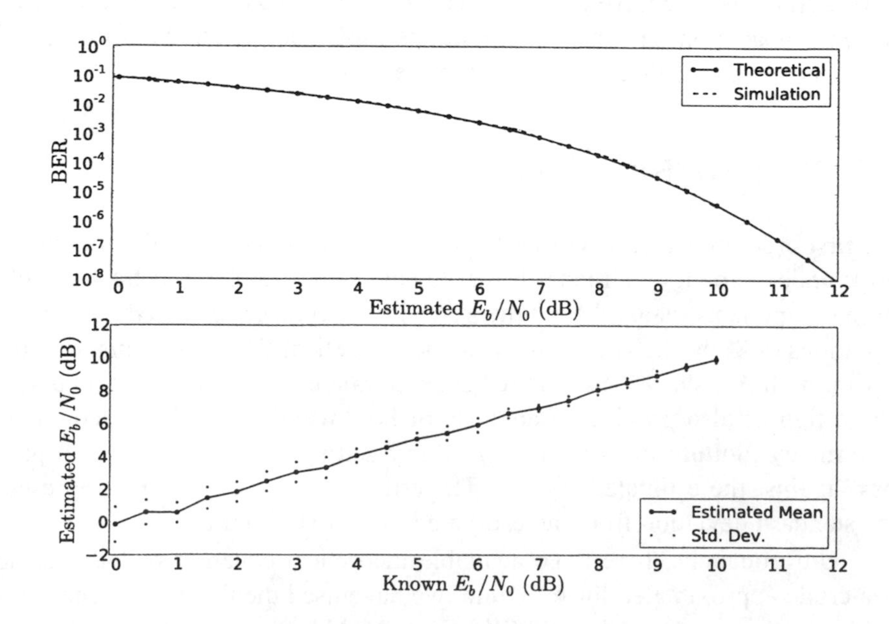

Figure A.2 BER curves and E_b/N_0 plots for QPSK.

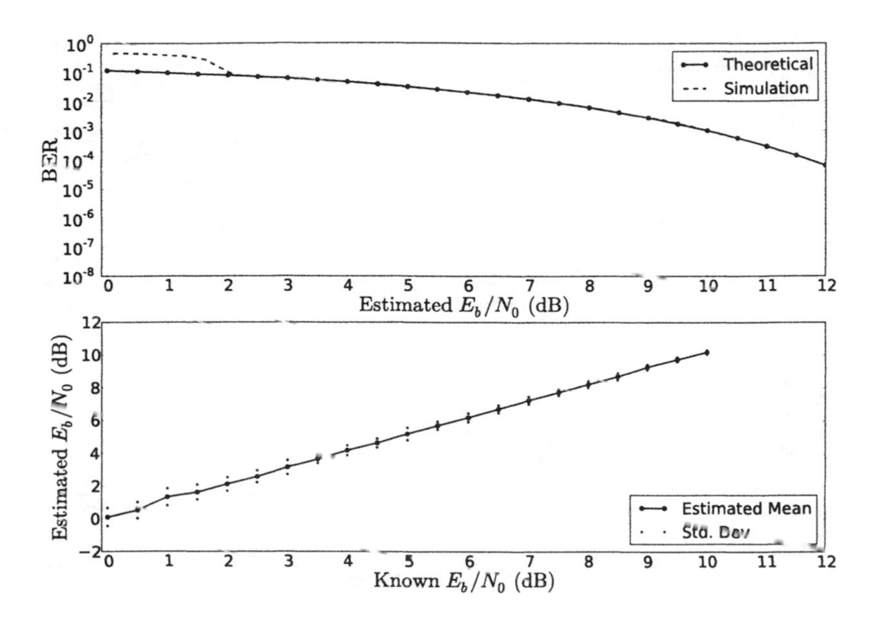

Figure A.3 BER curves and E_b/N_0 plots for 8PSK.

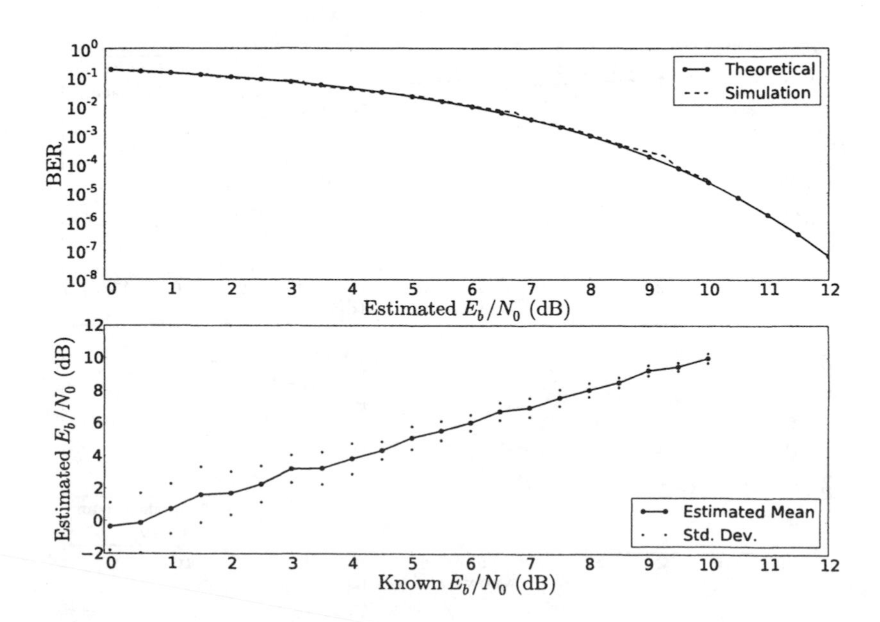

Figure A.4 BER curves and E_b/N_0 plots for DBPSK.

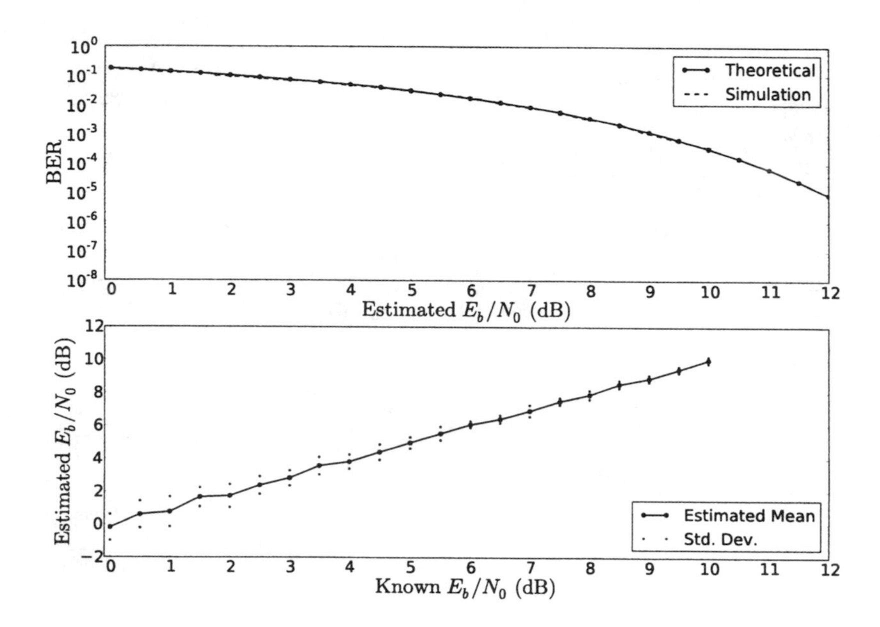

Figure A.5 BER curves and E_b/N_0 plots for DQPSK.

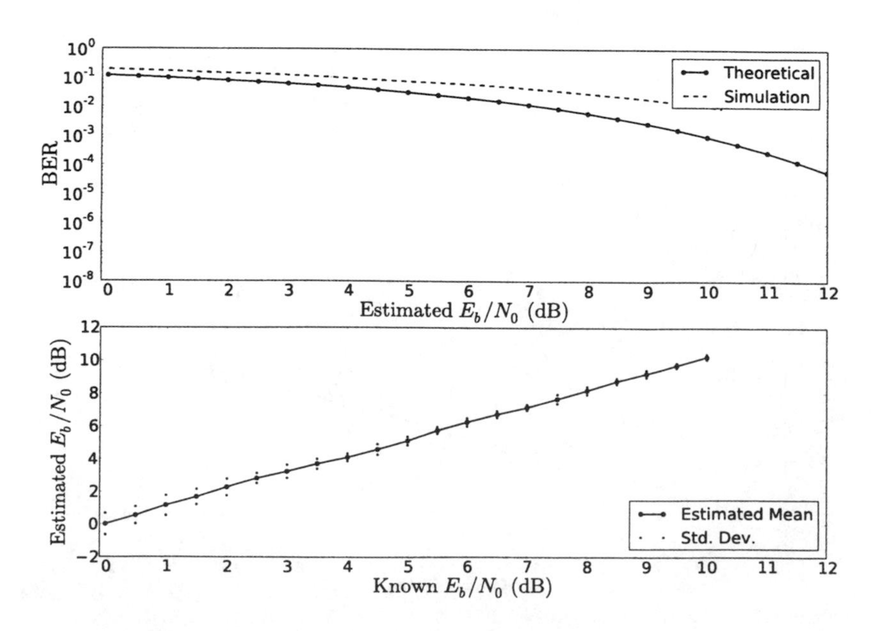

Figure A.6 BER curves and E_b/N_0 plots for D8PSK.

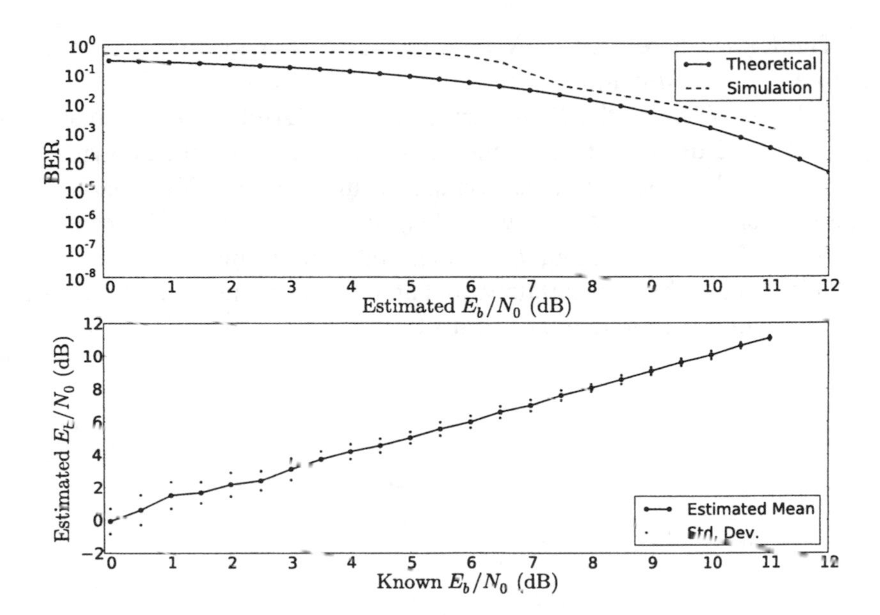

Figure A.7 BER curves and E_b/N_0 plots for GMSK.

The BER simulations for BPSK, QPSK, DBPSK, and DQPSK all matched very well to the theoretical curves. The 8PSK curve matches well for E_b/N_0 above 2 dB. For low E_b/N_0 values, it is possible that the synchronization loops or some other part of the receiver chain failed. The D8PSK curve looks consistent over the plotted E_b/N_0, but without a reference curve, it is difficult to say how well it compares to theory. The plotted curve in Figure A.6 shows the D8PSK simulated curve is about 2 to 3 dB worse in performance than the theoretical 8PSK curve, which is about the correct loss between differential and nondifferential modulations. The GMSK curve performs like the 8PSK curve where the system has poor performance under low E_b/N_0 conditions, although this behavior lasts up to about 7.5 dB. After that, the simulated curve is about 1 dB worse than the theoretical curve.

It is not the purpose of this analysis to provide the best performance for each of these modulation schemes. In this case, we are simply interested in the known baseline performance for the system under test. The performance provides an interesting problem for the combined optimization and learning system. The optimization routine assumes the theoretical performance of the given waveforms, so the learning system has to understand the difference between the estimated performance and the actual performance to help the optimizer to make better, more educated choices.

The figures here also show that the signal strength, noise power, and therefore E_b/N_0 estimations behave very well, especially at E_b/N_0 values over 4 dB. Looking at the BER curves for, say, DBPSK, the estimation at E_b/N_0 of 1 to 2 dB are off where the estimated average for the known 1.5 dB SNR is lower than the estimated average for the 1 dB E_b/N_0 case. Looking at the standard deviations, at low E_b/N_0, the estimated E_b/N_0 can be off by about 1 dB. Estimations of E_b/N_0 and other performance metrics at low E_b/N_0 is going to have uncertainty, and ambiguity and uncertainty of data is a problem which a cognitive radio must tolerate.

B

Additional BER Formulas

Section 4.2.1 provided the basic BER formulas for the modulation types used in the simulations and experiments in AWGN channels. Here, we present a few more BER formulas, including M-QAM in AWGN and other modulations in fading channels.

M-QAM in AWGN:

In this equation, γ_b represents the energy per bit.

$$P_e = \left(\frac{2}{\log_2(M)}\right)\left(\frac{\sqrt{M}-1}{\sqrt{M}}\right) erfc\left(\sqrt{\frac{3\log_2 M}{2(M-1)}\gamma_b}\right) \tag{B.1}$$

In fading channels, [1] provides the closed-form solutions for different modulations. Each formula comes from the basic equation for the probability of a symbol error in (B.2).

$$P_e = \int_0^\infty P_{AWGN}(x)p(x)dx \tag{B.2}$$

Where $p(x)$ is the probability density function (PDF) of the channel.

The closed form solutions to M-PSK and M-QAM modulations are defined in the following equations where $I(\bar{\gamma}, g, \theta)$ is specific to the channel; $\bar{\gamma}$ is the average signal to noise ratio, g is a modulation coefficient, and θ is the variable of integration. These formulas are modified from [1] to a single channel receiver instead of the L-finger maximal ratio combining (MRC) receiver.

BPSK:

$$\begin{aligned} P_e =& \frac{1}{\pi} \int_0^{\frac{\pi}{2}} I(\bar{\gamma}, g, \theta) d\theta \\ g =& 1 \end{aligned} \tag{B.3}$$

M-PSK:

$$\begin{aligned} P_e =& \frac{1}{\pi} \int_0^{\frac{(M-1)\pi}{M}} I(\bar{\gamma}, g, \theta) d\theta \\ g =& \sin^2\left(\frac{\pi}{M}\right) \end{aligned} \tag{B.4}$$

M-QAM:

$$\begin{aligned} P_e =& \frac{4}{\pi}\left(1 - \frac{1}{\sqrt{M}}\right) \int_0^{\frac{\pi}{2}} I(\bar{\gamma}, g, \theta) d\theta - \\ & \frac{4}{\pi}\left(1 - \frac{1}{\sqrt{M}}\right)^2 \int_0^{\frac{\pi}{4}} I(\bar{\gamma}, g, \theta) d\theta \\ g =& \frac{3}{2(M-1)} \end{aligned} \tag{B.5}$$

Rayleigh channels:

$$p(\gamma; \bar{\gamma}) = \frac{1}{\bar{\gamma}} \exp\left(-\frac{\gamma}{\bar{\gamma}}\right) \tag{B.6}$$

$$I(\bar{\gamma}, g, \theta) = \left(1 + \frac{g\bar{\gamma}}{\sin^2\theta}\right)^{-1} \tag{B.7}$$

Ricean channels:

$$\begin{aligned} p(\gamma; \bar{\gamma}, n) =& \frac{(1+n^2)\, e^{-n^2}}{\bar{\gamma}} \exp\left(-\frac{(1+n^2)\,\gamma}{\bar{\gamma}}\right) \times \\ & I_0\left(2n\sqrt{\frac{(1+n^2)\,\gamma}{\bar{\gamma}}}\right) \end{aligned} \tag{B.8}$$

where n^2 = Ricean factor

$$I(\bar{\gamma}, g, \theta) = \left(\frac{(1+n^2)\sin^2\theta}{(1+n^2)\sin^2\theta + g\bar{\gamma}}\right) \exp\left(\frac{n^2 g\bar{\gamma}}{(1+n^2)\sin^2\theta + g\bar{\gamma}}\right) \tag{B.9}$$

Nakagami-m channels:

$$p(\gamma;\bar{\gamma},m) = \frac{m^m \gamma^{m-1}}{\gamma^{-m}\Gamma(m)} \exp\left(-\frac{m\gamma}{\bar{\gamma}}\right) \tag{B.10}$$

where m = 1/2 for one-sided Gaussian

where m = 1 for Rayleigh channel

where $m = \infty$ for no fading

$\Gamma(m)$ is the gamma function

$$I(\bar{\gamma},g,\theta) = \left(\frac{(1+n^2)\sin^2\theta}{(1+n^2)\sin^2\theta + g\bar{\gamma}}\right) \exp\left(\frac{n^2 g\bar{\gamma}}{(1+n^2)\sin^2\theta + g\bar{\gamma}}\right) \tag{B.11}$$

References

[1] M. K. Simon and M. Alouini, "A Unified Approach to the Performance Analysis of Digital Communication over Generalized Fading Channels," *Proc. IEEE*, Vol. 86, No. 9, pp. 1860 – 1877, Sep. 1998.

C

OProfile and Results of Profiling GNU Radio

C.1 Introduction to *OProfile*

OProfile is an open source, GPL tool for measuring software performance [1]. Built specifically for Linux, *OProfile* uses the Linux kernel to read the processor's hardware counters as a measure of software complexity. It has low overhead on the system and resides as a separate process to monitor the performance of the entire system, which means a developer does not have to add specific hooks to enable profiling as other profilers require. As the website points out, the profiler monitors all system activity including "hardware and software interrupt handlers, kernel modules, the kernel, shared libraries, and applications." More importantly to application and library developers, the profiler keeps track of performance per symbol of each process, which means a developer can analyze the performance of individual functions and routines within the code. We use this last feature to understand the complexity of each of the modulators used in the GNU Radio.

C.2 *OProfile* Results of GNU Radio Modulators

Each block in a GNU Radio flow graph is a class in the GNU Radio library. Each class has a few callable functions, most important of which is the *work* function that performs the core of the signal processing. The interest in the profiling is discussed in Section 4.2.8 where the optimization process uses the computational complexity. In the analysis, the only blocks that change in the GNU Radio flow graph with the waveforms are the blocks that make up the modulators and demodulators. The rest of the blocks remain the same with different parameters; however, the computational performance remains the same when changing the transmitter power or carrier frequency. The only other performance difference results in changes in the symbol rate, which has an overall affect on each block as the sampling rate changes. Therefore, the performance profiling only looks at the blocks that make up the modulators and demodulators.

The performance analysis consists of running a GNU Radio simulation with just the modulator or demodulator and the required sinks and sources to pass a set number of symbols at the same symbol rate through the flow graph while running *OProfile*. Figure C.1(a) shows the GNU Radio flow graph of the modulator simulation and Figure C.1(b) shows the same for the demodulators. The profiler collects the number of ticks of the processor hardware counter during the set number of symbols to measure how much time each modulator and demodulator uses. Table C.1 shows an example output of the profiler.

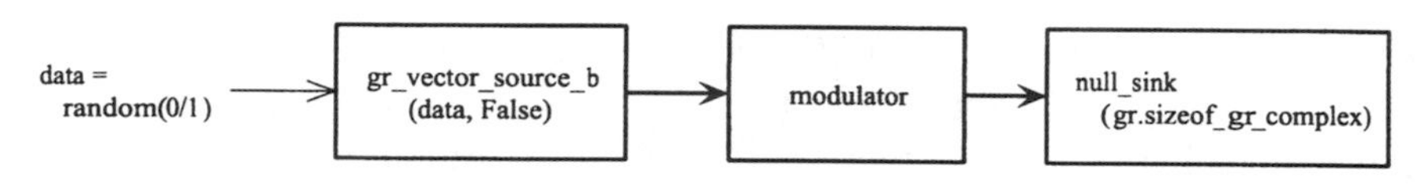

(a) Modulators take in bits and produce complex symbols.

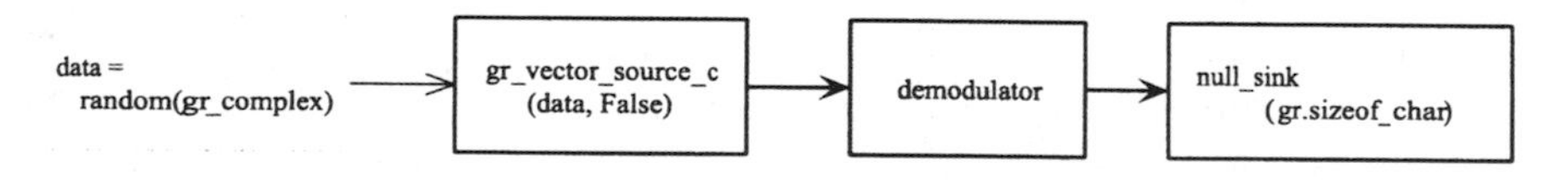

(b) Demodulators take in complex symbols and produce bits.

Figure C.1 Flow graphs for profiling GNU Radio (a) modulators and (b) demodulators.

Notice that each time a simulation is run, the performance counters had slight variations despite running with the same code, same number of symbols, and on the same platform. The system this was run on also runs other programs along with the profiler and GNU Radio. Furthermore, there are loops and branches in the software that depend on the value of the data, which is influenced by the random noise of the channel and will affect the performance. These issues together with other operating system factors are the probable causes of any variations among profiling runs. These should be statistically insignificant, and so each simulation is run ten times and averaged.

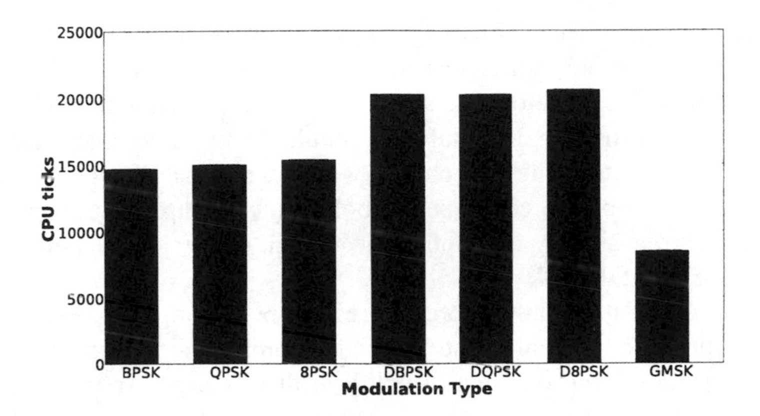

Figure C.2 Performance comparison of the available GNU Radio modulators with *OProfile*.

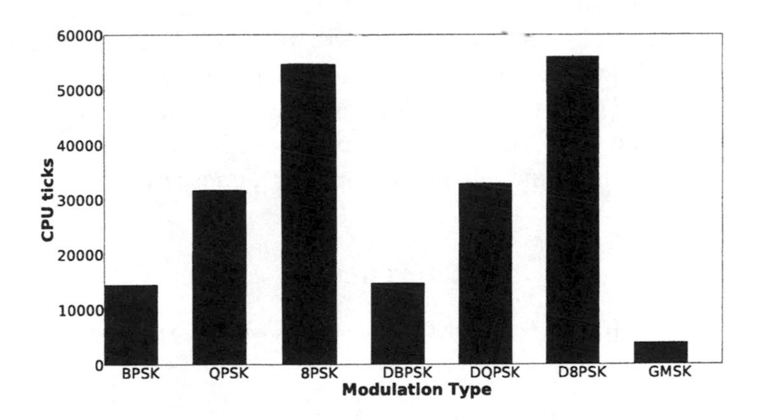

Figure C.3 Performance comparison of the available GNU Radio demodulators with *OProfile*.

Figure C.2 shows the resulting complexity of the modulators and Figure C.3 shows the complexity of the demodulators.

These complexity graphs are plotted with the one standard deviation error bars, which show that the random performance counter values are indeed insignificant. For the modulators, GMSK shows the smallest computational footprint in both modulator and demodulator, and the other modulators and demodulators shows definite trends related to their implementation. GMSK is implemented very simply, and the demodulator does not use an AGC loop. The M-PSK modulations all show increasing complexity with M, with a slight dip in the complexity of the DQPSK over DBPSK. In the implementation, the only difference between the differential modulators and the nondifferential modulators is a single block that performs the differential mapping, so the differential modulators should be more complex. Between the different modulators, the only change when increasing M is the block that maps the $\log_2(M)$ bits to complex symbols via a lookup table. Apparently, the lookup table is slightly more efficient, though hardly significantly, in the DQPSK case than DBPSK.

The demodulators show more interesting behavior. The demapping is done by brute-force matching the incoming complex signal to the nearest point by calculating the minimum Euclidean distance. As M increases, the performance also increases as the loop has more points to test. Like the modulator, the only difference between the differential and nondifferential cases is the use of a differential phasor block to decode the symbols. Although there are more efficient receivers that do not rely on the same complex synchronization loops for differential modulations, the receiver implementation in GNU Radio does not currently take advantage of them.

Table C.2 shows the same information in table format similar to the database used in the cognitive engine's optimization calculation. In this table, the modulators and demodulators are mixed together to form the overall complexity of selecting a modulation scheme for a transceiver.

References

[1] "OProfile," 2007. [Online]. Available: *http://oprofile.sourceforge.net/news/*

Table C.1
OProfile Results of DBPSK Modulator

Samples	%	Symbol name
13815	28.5841	.loop1
9895	20.4734	gr_fir_ccf_simd::filter(complex<float> const*)
7818	16.1760	.loop2
4139	8.5639	gr_interp_fir_filter_ccf::work(int, vector<void const*, allocator<void const*>>&, vector<void*>&)
3950	8.1728	gr_chunks_to_symbols_bc::work(int, vector<void const*, allocator<void const*>>&, vector<void*>&)
3265	6.7555	.cleanup
2009	4.1568	gr_diff_encoder_bb::work(int, vector<void const*, allocator<void const*>>&, vector<void*>&)
1151	2.3815	fcomplex_dotprod_sse
604	1.2497	gr_packed_to_unpacked_bb::general_work (int, vector<int>&, vector<void const*, allocator<void const*>>&, vector<void*>&)
435	0.9000	.plt
380	0.7862	get_bit_be(unsigned char const*, unsigned int)
293	0.6062	gr_single_threaded_scheduler::main_loop()
189	0.3911	gr_map_bb::work(int, vector<void const*, allocator<void const*>>&, vector<void*>&)
98	0.2028	gr_block_detail::input(unsigned int)
58	0.1200	min_available_space(gr_block_detail*, int)
40	0.0828	gr_vector_source_b::work(int, vector<void const*, allocator<void const*>>&, vector<void*>&)
29	0.0600	gr_buffer_reader::items_available() const
21	0.0435	gr_packed_to_unpacked_bb::forecast(int, vector<int>&)
18	0.0372	gr_buffer::space_available() const
17	0.0352	gr_sync_interpolator::general_work(int, vector<int>&, vector<void const*, allocator<void const*>>&, vector<void*>&)
12	0.0248	vector<int>::_M_fill_insert(vector<int>::iterator, unsigned long, int const&)

Samples	%	Symbol name
11	0.0228	gr_buffer::write_pointer()
11	0.0228	gr_sync_block::fixed_rate_ninput _to_noutput(int)
8	0.0166	gr_block_detail::produce_each(int)
8	0.0166	gr_sync_block::forecast(int, vector<int>&)
8	0.0166	vector<void*>::_M_fill_insert (vector<void*>::iterator, unsigned long, void* const&)
7	0.0145	gr_buffer_reader::read_pointer()
7	0.0145	gr_sync_interpolator::fixed_rate _noutput _to_ninput(int)
5	0.0103	gr_block_detail::consume_each(int)
5	0.0103	gr_buffer::update_write_pointer(int)
5	0.0103	gr_sync_block::general_work(int, vector<int>&, vector<void const*, allocator<void const*>>&, vector<void*>&)
5	0.0103	gr_sync_interpolator::forecast(int, vector<int>&)
3	0.0062	gr_sync_block::fixed_rate_noutput _to_ninput(int)
3	0.0062	vector<void const*, allocator<void const*>>:: _M_fill_insert(__gnu_cxx::__normal _iterator<void const**, vector<void const*, allocator<void const*>>, unsigned long, void const* const&)
3	0.0062	vector<void*>::erase(vector<void*>:: iterator, vector<void*>::iterator)
1	0.0021	global constructors keyed to _ZN8gr_prefs9singletonEv
1	0.0021	gr_block:: gr_block()
1	0.0021	gr_buffer_add_reader(boost:: shared_ptr<gr_buffer>, int)
1	0.0021	gr_buffer_reader::update_read_pointer(int)
1	0.0021	gr_prefix()
1	0.0021	void fill<vector<void*>::iterator, void*> (vector<void*>::iterator, vector<void*>::iterator, void* const&)

Table C.2
Computational Database for GNU Radio Modulators

ID	Modulation	Hardware Counter
1	BPSK	29142.8
2	QPSK	46615.0
3	8PSK	69998.0
4	DBPSK	34889.3
5	DQPSK	52996.5
6	D8PSK	76539.3
7	GMSK	12352.8

XML and DTD Representation of the Cognitive Components

This appendix provides template DTD and XML files used to pass information between the cognitive components. The DTD files are used by the cognitive engine to teach the cognitive controller the format the data it receives will look like and therefore how to process and store the data. The DTD files are transferred from the components to the cognitive controller during the initialization stage. The waveform file representation is used in the genetic algorithm to understand how to build the chromosome to properly represent the system.

D.1 Waveform Representation

gnuradio_lb.dtd

```
<!ELEMENT waveform              (Tx,Rx)>
<!ATTLIST waveform type         #CDATA "analog/digital">
<!ELEMENT Tx                    (PHY,LINK)>
<!ELEMENT PHY                   (rf,mod)>
<!ELEMENT rf                    (tx_freq+,tx_power+)>
<!ELEMENT tx_freq               (min,max,step)>
<!ELEMENT min                   (#PCDATA)>
<!ELEMENT max                   (#PCDATA)>
<!ELEMENT step                  (#PCDATA)>
<!ELEMENT tx_power              (min,max,step)>
<!ELEMENT min                   (#PCDATA)>
<!ELEMENT max                   (#PCDATA)>
<!ELEMENT step                  (#PCDATA)>
<!ELEMENT mod                   (tx_mod+,tx_rolloff?,tx_bt?,
                                tx_gray_code?,tx_symbol_rate)>
<!ELEMENT tx_mod                (tx_mod_bits,tx_mod_differential)>
<!ATTLIST tx_mod type (psk,msk,qam) "psk">
```

```
<!ELEMENT tx_mod_bits          (min,max,step)>
<!ELEMENT min                  (#PCDATA)>
<!ELEMENT max                  (#PCDATA)>
<!ELEMENT step                 (#PCDATA)>
<!ELEMENT tx_mod_differential  (min,max,step)>
<!ELEMENT min                  (#PCDATA)>
<!ELEMENT max                  (#PCDATA)>
<!ELEMENT step                 (#PCDATA)>
<!ELEMENT tx_rolloff           (min,max,step)>
<!ELEMENT min                  (#PCDATA)>
<!ELEMENT max                  (#PCDATA)>
<!ELEMENT step                 (#PCDATA)>
<!ELEMENT tx_bt                (min,max,step)>
<!ELEMENT min                  (#PCDATA)>
<!ELEMENT max                  (#PCDATA)>
<!ELEMENT step                 (#PCDATA)>
<!ELEMENT tx_gray_code         (min,max,step)>
<!ELEMENT min                  (#PCDATA)>
<!ELEMENT max                  (#PCDATA)>
<!ELEMENT step                 (#PCDATA)>
<!ELEMENT tx_symbol_rate       (min,max,step)>
<!ELEMENT min                  (#PCDATA)>
<!ELEMENT max                  (#PCDATA)>
<!ELEMENT step                 (#PCDATA)>
<!ELEMENT LINK                 (frame)>
<!ELEMENT frame                (tx_pkt_size,tx_access_code?)>
<!ELEMENT tx_pkt_size          (min,max,step)>
<!ELEMENT min                  (#PCDATA)>
<!ELEMENT max                  (#PCDATA)>
<!ELEMENT step                 (#PCDATA)>
<!ELEMENT tx_access_code       (min,max,step)>
<!ELEMENT min                  (#PCDATA)>
<!ELEMENT max                  (#PCDATA)>
<!ELEMENT step                 (#PCDATA)>
<!ELEMENT Rx                   (PHY,LINK)>
<!ELEMENT PHY                  (rf+,mod,frame_correlator)>
<!ELEMENT rf                   (rx_freq+,rx_gain+)>
<!ELEMENT rx_freq              (min,max,step)>
<!ELEMENT min                  (#PCDATA)>
<!ELEMENT max                  (#PCDATA)>
<!ELEMENT step                 (#PCDATA)>
<!ELEMENT rx_gain              (min,max,step)>
<!ELEMENT min                  (#PCDATA)>
<!ELEMENT max                  (#PCDATA)>
<!ELEMENT step                 (#PCDATA)>
<!ELEMENT mod                  (rx_mod+,rx_rolloff?,rx_bt?,
                               rx_gray_code?,rx_symbol_rate)>
<!ELEMENT rx_mod               (rx_mod_bits,rx_mod_differential)>
<!ATTLIST rx_mod type (psk,msk,qam) "psk">
<!ELEMENT rx_mod_bits          (min,max,step)>
<!ELEMENT min                  (#PCDATA)>
<!ELEMENT max                  (#PCDATA)>
<!ELEMENT step                 (#PCDATA)>
<!ELEMENT rx_mod_differential  (min,max,step)>
<!ELEMENT min                  (#PCDATA)>
<!ELEMENT max                  (#PCDATA)>
```

```
<!ELEMENT step                (#PCDATA)>
<!ELEMENT rx_rolloff          (min,max,step)>
<!ELEMENT min                 (#PCDATA)>
<!ELEMENT max                 (#PCDATA)>
<!ELEMENT step                (#PCDATA)>
<!ELEMENT rx_bt               (min,max,step)>
<!ELEMENT min                 (#PCDATA)>
<!ELEMENT max                 (#PCDATA)>
<!ELEMENT step                (#PCDATA)>
<!ELEMENT rx_gray_code        (min,max,step)>
<!ELEMENT min                 (#PCDATA)>
<!ELEMENT max                 (#PCDATA)>
<!ELEMENT step                (#PCDATA)>
<!ELEMENT rx_symbol_rate      (min,max,step)>
<!ELEMENT min                 (#PCDATA)>
<!ELEMENT max                 (#PCDATA)>
<!ELEMENT step                (#PCDATA)>
<!ELEMENT frame_correlator    (ACthreshold)>
<!ELEMENT ACthreshold         (min,max,step)>
<!ELEMENT min                 (#PCDATA)>
<!ELEMENT max                 (#PCDATA)>
<!ELEMENT step                (#PCDATA)>
<!ELEMENT LINK                (frame)>
<!ELEMENT frame               (rx_pkt_size,rx_access_code?)>
<!ELEMENT rx_pkt_size         (min,max,step)>
<!ELEMENT min                 (#PCDATA)>
<!ELEMENT max                 (#PCDATA)>
<!ELEMENT step                (#PCDATA)>
<!ELEMENT rx_access_code      (min,max,step)>
<!ELEMENT min                 (#PCDATA)>
<!ELEMENT max                 (#PCDATA)>
<!ELEMENT step                (#PCDATA)>
```

gnuradio_lb.xml

```
<?xml version=“1.0”?>
<!DOCTYPE WAVEFORM SYSTEM “gnuradio_lb.dtd”>
<waveform type=“digital”>
  <Tx>
    <PHY>
      <rf>
        <tx_freq>
          <min unit=“kHz”>400000<\min>
          <max unit=“kHz”>500000<\max>
          <step unit=“kHz”>1<\step>
        <\tx_freq>
        <tx_power>
          <min unit=“dBm”>0<\min>
          <max unit=“dBm”>100<\max>
          <step unit=“dBm”>0.1<\step>
        <\tx_power>
      <\rf>
      <rf>
        <tx_freq>
          <min unit=“kHz”>2300000<\min>
```

```
      <max unit="kHz">2500000< \max>
      <step unit="kHz">100< \step>
    < \tx_freq>
    <tx_power>
      <min unit="dBm">0< \min>
      <max unit="dBm">20< \max>
      <step unit="dBm">0.1< \step>
    < \tx_power>
  < \rf>
  <mod>
    <tx_mod type="PSK">
      <tx_mod_bits>
          <min>1< \min>
          <max>3< \max>
          <step>1< \step>
      < \tx_mod_bits>
      <tx_mod_differential>
          <min>0< \min>
          <max>1< \max>
          <step>1< \step>
      < \tx_mod_differential>
    < \tx_mod>
    <tx_mod type="GMSK">
      <tx_mod_bits>
          <min>1< \min>
          <max>1< \max>
          <step>1< \step>
      < \tx_mod_bits>
      <tx_mod_differential>
          <min>0< \min>
          <max>0< \max>
          <step>0< \step>
      < \tx_mod_differential>
    < \tx_mod>
    <tx_rolloff units="na">
      <min>0< \min>
      <max>1< \max>
      <step>0.01< \step>
    < \tx_rolloff>
    <tx_bt units="na">
      <min>0< \min>
      <max>1< \max>
      <step>0.01< \step>
    < \tx_bt>
    <tx_gray_code>
      <min>0< \min>
      <max>1< \max>
      <step>1< \step>
    < \tx_gray_code>
    <tx_symbol_rate units="Hz" mult="1">
      <min>0.1< \min>
      <max>1.0< \max>
      <step>0.125< \step>
    < \tx_symbol_rate>
  < \mod>
< \PHY>
```

```
    <LINK>
      <frame>
        <tx_pkt_size units="bytes">
          <min>1< \min>
          <max>1500< \max>
          <step>1< \step>
        < \tx_pkt_size>
        <tx_access_code>None< \tx_access_code>
      < \frame>
    < \LINK>
  < \Tx>
< \waveform>
```

D.2 Objectives Sensor

sensor_objectives.dtd

```
<!ELEMENT sensor                         (awgn,fer,powerconsumption,
                                         sinr,throughput,
                                         spectralefficiency,
                                         bandwidth,interference,
                                         computationalcomplexity)>
<!ATTLIST sensor                         name #CDATA #REQUIRED>
<!ELEMENT awgn                           (#PCDATA)>
<!ATTLIST awgn                           type (float) "float">
<!ATTLIST awgn                           typeref (phy,mac,sys) "phy">
<!ELEMENT fer                            (#PCDATA)>
<!ATTLIST fer                            type (float) "float">
<!ATTLIST fer                            typeref (phy,mac,sys) "mac">
<!ELEMENT sinr                           (#PCDATA)>
<!ATTLIST sinr                           type (float) "float">
<!ATTLIST sinr                           typeref (phy,mac,sys) "phy">
<!ELEMENT throughput                     (#PCDATA)>
<!ATTLIST throughput                     type (float) "float">
<!ATTLIST throughput                     typeref (phy,mac,sys) "phy">
<!ELEMENT bandwidth                      (#PCDATA)>
<!ATTLIST bandwidth                      type (float) "float">
<!ATTLIST bandwidth                      typeref (phy,mac,sys) "phy">
<!ELEMENT spectralefficiency             (#PCDATA)>
<!ATTLIST spectralefficiency             type (float) "float">
<!ATTLIST spectralefficiency             typeref (phy,mac,sys) "phy">
<!ELEMENT interference                   (#PCDATA)>
<!ATTLIST interference                   type (float) "float">
<!ATTLIST interference                   typeref (phy,mac,sys) "phy">
<!ELEMENT powerconsumption               (#PCDATA)>
<!ATTLIST powerconsumption               type (float) "float">
<!ATTLIST powerconsumption               typeref (phy,mac,sys) "sys">
<!ELEMENT computationalcomplexity        (#PCDATA)>
<!ATTLIST computationalcomplexity        type (float) "float">
<!ATTLIST computationalcomplexity        typeref (phy,mac,sys) "sys">
```

sensor_objectives.xml

```
<?xml version="1.0"?>
```

```
<!DOCTYPE sensor SYSTEM "sensor_objectives.dtd">
<sensor name="objectives">
  <awgn type="float" typeref="phy">0.0<\awgn>
  <fer type="float" typeref="phy">0.0<\fer>
  <sinr type="float" typeref="phy">0.0<\sinr>
  <throughput type="float" typeref="phy">0.0<\throughput>
  <bandwidth type="float" typeref="phy">0.0<\bandwidth>
  <spectralefficiencytype="float" typeref="phy">0.0
                      <\spectralefficiency>
  <interference type="float" typeref="phy">0.0<\interference>
  <powerconsumption type="float" typeref="phy">0.0
                      <\powerconsumption>
  <computationalcomplexity type="float" typeref="phy">0.0
                      <\computationalcomplexity>
<\sensor>
```

D.3 Meters Sensor

sensor_meters.dtd

```
<!ELEMENT sensor              (ber,per,ebno,tx_signal_power,
                              rx_signal_power,noise_power)>
<!ATTLIST sensor              name #CDATA #REQUIRED>
<!ELEMENT ber                 (#PCDATA)>
<!ATTLIST ber                 type (float) "float">
<!ATTLIST ber                 size #CDATA "1">
<!ELEMENT per                 (#PCDATA)>
<!ATTLIST per                 type (float) "float">
<!ATTLIST per                 size #CDATA "1">
<!ELEMENT ebno                (#PCDATA)>
<!ATTLIST ebno                type (float) "float">
<!ATTLIST ebno                size #CDATA "1">
<!ELEMENT tx_signal_power     (#PCDATA)>
<!ATTLIST tx_signal_power     type (float) "float">
<!ATTLIST tx_signal_power     size #CDATA "1">
<!ELEMENT rx_signal_power     (#PCDATA)>
<!ATTLIST rx_signal_power     type (float) "float">
<!ATTLIST rx_signal_power     size #CDATA "1">
<!ELEMENT noise_power         (#PCDATA)>
<!ATTLIST noise_power         type (float) "float">
<!ATTLIST noise_power         size #CDATA "1">
```

sensor_meters.xml

```
<?xml version="1.0"?>
<!DOCTYPE sensor SYSTEM "sensor_meters.dtd">
<sensor name="meters">
  <ber type="float" size="1">0<\ber>
  <per type="float" size="1">0<\per>
  <ebno type="float" size="1" units="dB">0<\ebno>
  <tx_signal_powertype="float" size="1" units="dBm">0
                 <\tx_signal_power>
  <rx_signal_power type="float" size="1" units="dBm">0
```

```
                    < \rx_signal_power>
  <noise_power type="float" size="1" units="dBm">0< \noise_power>
< \sensor>
```

D.4 PSD Sensor

sensor_psd.dtd

```
<!ELEMENT sensor              (noise_floor,signal*)>
<!ATTLIST sensor              name #CDATA #REQUIRED>
<!ELEMENT noise_floor         (#PCDATA)>
<!ATTLIST noise_floor         type (float) "float">
<!ATTLIST noise_floor         size #CDATA "1">
<!ELEMENT signal              (amplitude, fmin, fmax)>
<!ELEMENT amplitude           (#PCDATA)>
<!ATTLIST amplitude           type (float) "float">
<!ATTLIST amplitude           size #CDATA "1">
<!ELEMENT fmin                (#PCDATA)>
<!ATTLIST fmin                type (float) "float">
<!ATTLIST fmin                size #CDATA "1">
<!ELEMENT fmax                (#PCDATA)>
<!ATTLIST fmax                type (float) "float">
<!ATTLIST fmax                size #CDATA "1">
```

sensor_psd.xml

```
<?xml version="1.0"?>
<sensor name="psd">
  <noise-floor type="float" size="1" unit="dBm">-85< \noise-floor>
  <signal>
    <amplitude type="float" size="1" unit="dBm">-50< \amplitude>
    <fmin type="float" size="1" unit="Hz">449e6< \fmin>
    <fmax type="float" size="1" unit="Hz">451e6< \fmax>
  < \signal>
< \sensor>
```

D.5 Cognitive Controller Configuration

The cognitive controller configuration script lists the components attached to the controller as well as certain pieces of information used to interact with the component. Most of the components are distributed processes that the controller communicates with over a TCP socket. For each of these, the source and destination address (either an IP address or the domain name of the node) are listed along with the source and destination ports. The source port is generally left as 0 so the socket system can select a free port. The destination port must match the port number the component is listening to and is arbitrary.

Here, the 1024 range is used only as an illustrative example; port numbers 0 through 1023 are assigned for global use by certain protocols.

The *radio_node* sections are used to describe attached radios that the cognitive radio communicates with on the network to transmit waveform information. The configuration can list any number of these here with their address and port information. The knowledge base node is currently directly associated with the cognitive controller, but as a MySQL database, the access is performed over a TCP socket, too, where the hostname tells the controller on which host the database is served along with the username and password credentials to access the database. This is a work in progress as the authentication information should be more secure.

cognitive_controller.xml

```
<?xml version="1.0"?>
<cognitive-controller>
   <defaults>
      <xmlscripts>../xmlscripts/< \xmlscripts>
      <waveform>default_sim_waveform.xml< \waveform>
   < \defaults>
   <knowledge-base>
      <name>casebase< \name>
      <hostname>localhost< \hostname>
      <dbsize>10< \dbsize>
      <nsolutions>1< \nsolutions>
   < \knowledge-base>
   <sensor>
      <name>psd< \name>
      <src_hostname>localhost< \src_hostname>
      <dst_hostname>localhost< \dst_hostname>
      <src_port>0< \src_port>
      <dst_port>1024< \dst_port>
   < \sensor>
   <sensor>
      <name>meters< \name>
      <src_hostname>localhost< \src_hostname>
      <dst_hostname>localhost< \dst_hostname>
      <src_port>0< \src_port>
      <dst_port>1025< \dst_port>
   < \sensor>
   <sensor>
      <name>objectives< \name>
      <src_hostname>localhost< \src_hostname>
      <dst_hostname>localhost< \dst_hostname>
      <src_port>0< \src_port>
      <dst_port>1026< \dst_port>
   < \sensor>
   <optimizer>
      <src_hostname>localhost< \src_hostname>
      <dst_hostname>localhost< \dst_hostname>
      <src_port>0< \src_port>
      <dst_port>1027< \dst_port>
```

```
    <parameters>parameters_wsga.xml< \parameters>
    <sdr_definition>gnuradio.xml< \sdr_definition>
  < \optimizer>
  <radio>
    <src_hostname>localhost< \src_hostname>
    <dst_hostname>localhost< \dst_hostname>
    <src_port>0< \src_port>
    <dst_port>1028< \dst_port>
    <radio_node>
        <src_hostname>localhost< \src_hostname>
        <dst_hostname>localhost< \dst_hostname>
        <control_port>1100< \control_port>
    < \radio_node>
  < \radio>
  <user-interface>
    <src_hostname>localhost< \src_hostname>
    <dst_hostname>localhost< \dst_hostname>
    <src_port>1029< \src_port>
    <dst_port>0< \dst_port>
  < \user-interface>
  <policy_engine>
    <src_hostname>localhost< \src_hostname>
    <dst_hostname>localhost< \dst_hostname>
    <src_port>0< \src_port>
    <dst_port>1030< \dst_port>
    <db_name>spectrum_mask< \db_name>
    <table_name>dyspan2007< \table_name>
    <username>*****< \username>
    <password>*****< \password>
  < \policy_engine>
< \cognitive-controller>
```

Optimal Solutions of Knapsack Problems

For the tests of the knapsack problems used with case-based decision theory work of Chapter 6, we randomly created a set of knapsack problems and stored these for repeated and comparable use. The results presented in Chapter 6 showed that many of the problems demonstrated significant improvement using CBDT while others did not perform as well. To further analyze this, we ran the simple knapsack GA for 50,000 generations to produce the optimal value (or at least very close to it). In Chapter 5, the knapsack GA showed asymptotic and therefore convergence behavior after 5,000 generations. We ran for an extra 10 times that many generations to improve this further. The results of this process are presented in Table E.1 for each of the 100 models used. Knowing this bound is useful because it helps understand how difficult the knapsack problem is to solve. Smaller overall profit values are much more difficult to solve than larger values. Two interesting models are 62 and 63, both with small overall profit values. The results of the CBDT are discussed in Chapter 6 where different methods provided different results; some produced significant improvement for model 62 but not for 63 while other methods produced significant improvement for model 63 and not 62. The results in this table show that these two problems are difficult to solve. When applying case base feedback, the initial solutions can either help find better solutions faster, or the initial solutions might bias the population to a local optimum and hurt the search for the global optimum. See Chapter 6 for more analysis of what these results mean.

Table E.1
Near-Optimal Values of Knapsack Models

Model	Near-Optimal Profit
model0	0.482983
model1	0.484326
model2	0.350086
model3	0.264022
model4	0.464853
model5	0.297343
model6	0.268765
model7	0.473757
model8	0.390390
model9	0.369584
model10	0.362265
model11	0.137403
model12	0.569386
model13	0.467299
model14	0.307893
model15	0.481234
model16	0.475427
model17	0.438981
model18	0.510267
model19	0.342045
model20	0.492449
model21	0.324600
model22	0.308674
model23	0.079252
model24	0.442939
model25	0.408664
model26	0.443296
model27	0.313626
model28	0.374601
model29	0.425843
model30	0.415269
model31	0.343033
model32	0.393600
model33	0.513806

Model	Near-Optimal Profit
model34	0.250659
model35	0.463317
model36	0.472347
model37	0.470804
model38	0.470514
model39	0.213563
model40	0.339058
model41	0.227216
model42	0.465122
model43	0.235960
model44	0.393179
model45	0.411411
model46	0.472914
model47	0.205941
model48	0.372415
model49	0.321990
model50	0.222235
model51	0.289213
model52	0.483975
model53	0.473257
model54	0.432361
model55	0.405035
model56	0.479364
model57	0.134768
model58	0.361675
model59	0.143135
model60	0.461016
model61	0.460595
model62	0.086188
model63	0.095681
model64	0.173661
model65	0.505214
model66	0.335363

Model	Near-Optimal Profit
model67	0.455881
model68	0.509552
model69	0.456414
model70	0.266002
model71	0.444824
model72	0.363761
model73	0.235444
model74	0.426592
model75	0.166180
model76	0.445178
model77	0.116757
model78	0.294269
model79	0.335633
model80	0.123189
model81	0.493576
model82	0.459684
model83	0.163396
model84	0.295574
model85	0.165456
model86	0.236255
model87	0.363123
model88	0.495544
model89	0.408303
model90	0.445354
model91	0.365454
model92	0.416054
model93	0.130179
model94	0.441690
model95	0.359674
model96	0.176658
model97	0.267482
model98	0.443524
model99	0.483214

F

Simulation of an SINR Sensor

F.1 Sensor Design

A transmitted signal, $s[k]$, contains a sequence of known data symbols, $s_k[k]$ such as a training sequence, preamble, or access code. The received signal, $r[k]$, is shown in (F.1) and includes the transmitted signal, interference signals in the channel bandwidth, and AWGN noise. Not represented in this equation is that the transmitted signals (including the interferers') amplitude is adjusted by some pathloss of the propagation channel.

$$r(t) = s[k] + \sum (i[k]) + n[k] \tag{F.1}$$

Correlating the received signal with the known data symbols will peak when the received signal is the known data sequence. The correlation can be implemented as an FIR filter the length of the known sequence and where the taps are the values of the known sequence. A similar autocorrelation is done using the known sequence to get the peak value for use in determining the pathloss. Equation (F.2) gives the peak value the cross-correlation where K is the length of the known sequence, and (F.3) gives the peak of the autocorrelation process.

$$m_x = \sum_{k=0}^{K-1} r[k] s_k[k] \tag{F.2}$$

$$m_a = \sum_{k=0}^{K-1} s_k[k] s_k[k] \tag{F.3}$$

The known sequence is then adjusted in amplitude for pathloss by the ratio m_x/m_a. The adjustment is based on the assumption that the pathloss does not significantly change over the duration of the known sequence, which is usually a few dozen bits long. The GNU Radio access code is 64 bits long. Subtracting the adjusted known sequence from the received signal leaves the interference plus noise (F.4). Subtracting the results of (F.4) from the received signal leaves just the transmitted signal (F.5). These equations are valid for k over the length of the known sequence, K, once the cross-correlation detects a received known sequence.

$$r_{n+i}[k] = r[k] - \frac{m_x}{m_a} s_k[k] = \sum (i[k]) + n[k] \tag{F.4}$$

$$r_s[k] = r[k] - r_{n+i}[k] \tag{F.5}$$

The signal power is then the average magnitude squared of r_s (F.6), and the interference plus noise power is the average magnitude squared of r_{n+i} (F.7).

$$s = \frac{1}{K} \sum_{k=0}^{K-1} |r_s[k]|^2 \tag{F.6}$$

$$i + n = \frac{1}{K} \sum_{k=0}^{K-1} |r_{i+n}[k]|^2 \tag{F.7}$$

Equation (F.8) is the estimated SINR.

$$SINR = 10 \log_{10} \left(\frac{s}{i+n} \right) \tag{F.8}$$

F.2 Simulation

The simulations were done using the MATLAB script provided below. The script requires the communications toolbox.

The first simulation looks at the SINR estimator without any interference, which should simply calculate the SNR. Figure F.1 plots the estimated results versus the known SNR. Each point is the average of 100 simulations using a 64-bit random known sequence. The figure shows a high precision in calculating the SNR. There is a constant offset of about 0.35 dB due to the use of root-raised cosine pulse shape filtering on previous symbols that is not subtracted out of the signal and thus adds a bit of extra energy to the calculation.

The purpose of this sensor is to calculate properly the signal power and the signal to interference plus noise power in the presence of interference. In

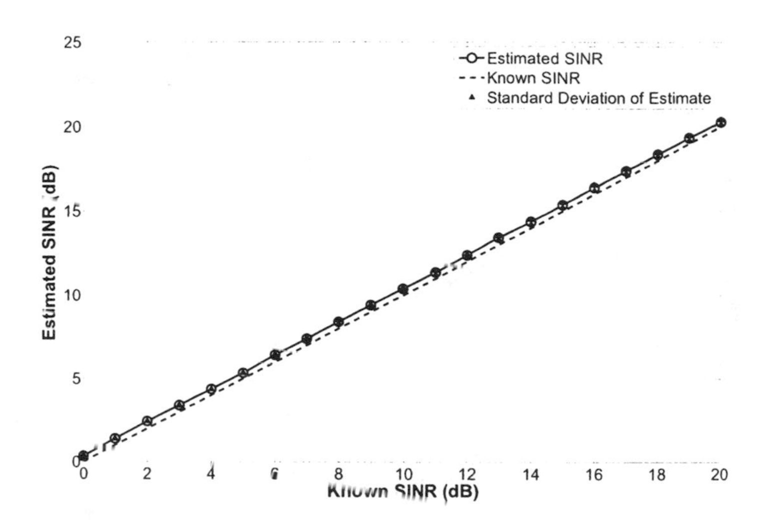

Figure F.1 Estimated SINR with no interference power.

the next simulation, both the signal and noise power are kept the same while the interference amplitude is adjusted from 0 to 1 V peak-to-peak. The SNR of the AWGN channel was set to 20 dB and the simulations are averaged 100 times for each interference amplitude setting. Figure F.2 shows the estimates of the SINR, which starts at 20 dB when the interference is not present and slopes downward to about 0 dB when the signal power of the interference is the same as the signal power of the transmitter. Furthermore, Figure F.3 shows the estimation of the signal power, which remains relatively constant for any power of interference. The standard deviation shows the estimates are within about 1 dB from the actual value when the interference power is at its highest. These simulation results show that this type of sensor will properly provide the cognitive engine with the required estimates of the signal, interference, and noise powers.

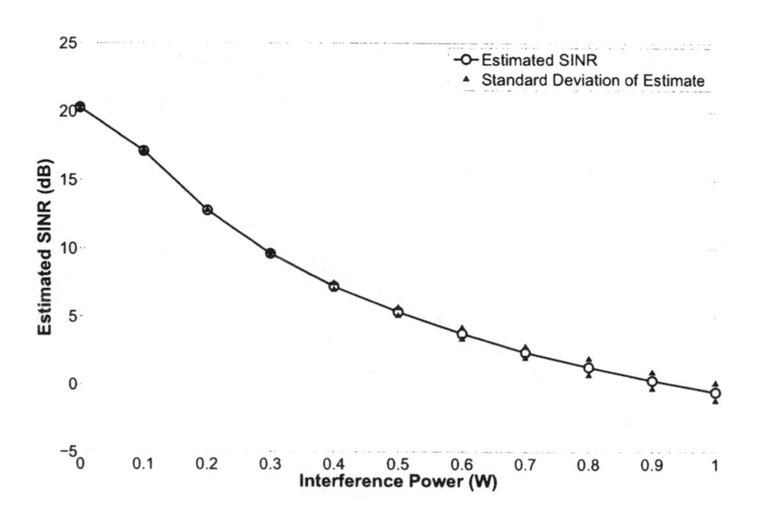

Figure F.2 Estimated SINR for varying amounts of interference.

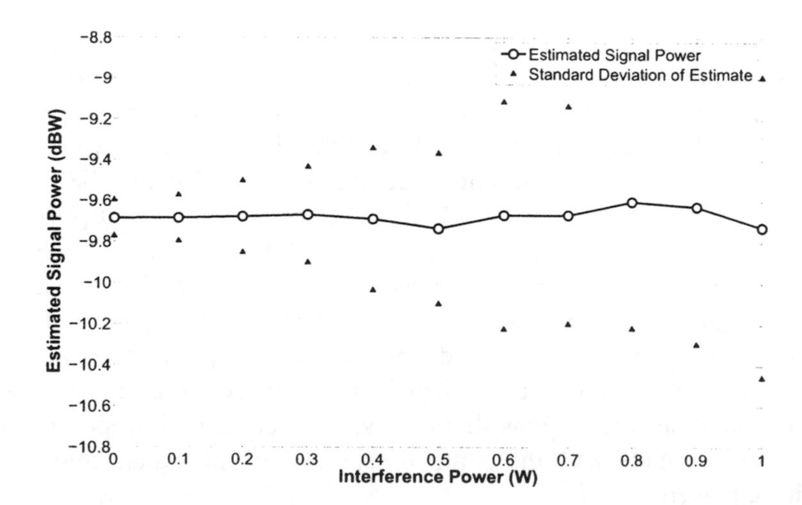

Figure F.3 Estimated received signal power for varying amounts of interference.

F.3 MATLAB Code

F.3.1 SINR Calculation Function

```
function [S,I,l] = corr_sir(SNR, iamp)
% Return the signal power (S) and interference plus noise power (I)
```

```
% and estimated pathloss (l) given a specified SNR (in dB) and amplitude
% of an interference signal

Ns = 1000;          % size of data
K = 64;             % size of known sequence (in bits)
sps = 8;            % number of samples per symbol
fd = 1;             % symbol rate
fs = sps*fd;        % sample rate
rc_alpha = 0.35;    % raised cosine filter rolloff factor

% Create training sequence
train = randint(1,K);
train_mod = pskmod(train, 2);
t = rcosflt(train_mod, fd, fs, 'fir/sqrt', rc_alpha);

% Create source signal including the training sequence
xs = [randint(1, Ns), train, randint(1, Ns)];
xs_mod = pskmod(xs, 2);
s = rcosflt(xs_mod, fd, fs, 'fir/sqrt', rc_alpha);

% Create interference signal
xi = randint(length(xs), 1);
xi_mod = pskmod(xi, 4);
i = iamp*rcosflt(xi_mod, fd, fs, 'fir/sqrt', 0.25);

% Received signal is sum of source, interference, and noise
r = awgn(s + i, SNR, 'measured');
r = 0.1*r; % crude pathloss

% Autocorrelate training sequence to get max correlation value
autocor = xcorr(t,t);
autocor = autocor(ceil(length(autocor)/2) : length(autocor));
[mauto, indauto] = max(autocor);

% Correlate against known training sequence and find peak
sigcor = xcorr(r, t);
sigcor = sigcor(ceil(length(sigcor)/2) : length(sigcor));
sigcor = abs(sigcor);
[m,ind] = max(sigcor);

% Use crosscorrelation and autocorrelation of training sequence
```

```
% to adjust the signal for differences in amplitude.
% In this simulation, this should be equal to the pathloss value
ratio = m / mauto;

% Remove training sequence to separate into (i+n) and (s)
a = 4.0; % remove outside edges to compensate for RRC
tadj = ratio * t(a*sps : length(t)-a*sps);
r_t = r(ind + a*sps - 1 : ind + a*sps + length(tadj) - 2);

r_i = r_t - tadj;
r_s = r_t - r_i;

% Calculate average magnitude squared to get signal and I+N power
S = 20*log10(mean(abs(r_s)));
I = 20*log10(mean(abs(r_i)));
l = ratio;
```

F.3.2 Plotting SINR with No Interference Power

```
clear
clc
n = 1;
sig = 0;
int = 0;
sinr = 0;
std_sig = 0;
std_int = 0;
std_sinr = 0;
for SNR = 0:1:20
        temp_sig = 0;
        temp_int = 0;
        for i = 1:100
            [S, I, l] = corr_sir(SNR, 0);
            temp_sig(i) = S;
            temp_int(i) = I;
        end
        sig(n) = mean(temp_sig);
        int(n) = mean(temp_int);
        sinr(n) = sig(n) - int(n);
```

```
        std_sig(n) = std(temp_sig);
        std_int(n) = std(temp_int);
        std_sinr(n) = std_sig(n) - std_int(n);

        n = n+1;
end

SNR = 0:1:20;
fig = figure(1);
plot(SNR, sinr, 'o-b', 'LineWidth', 2.0, 'MarkerFace', 'b', 'MarkerSize', 10)
hold on
plot(SNR, SNR, '-k', 'LineWidth', 2.0)
plot(SNR, sinr + std_sinr, '^k', 'MarkerFace', 'k')
plot(SNR, sinr - std_sinr, '^k', 'MarkerFace', 'k')

xlabel('Known SINR (dB)', 'FontSize', 20, 'FontWeight', 'bold');
ylabel('Estimated SINR (dB)', 'FontSize', 20, 'FontWeight', 'bold');
set(fig, 'Position', [100 100 1000 640])
set(gca, 'FontSize', 18);
set(gca, 'Position', [0.075 0.1 0.875 0.875])
leg = legend('Estimated SINR', 'Known SINR', 'Standard Deviation of Estimate');
```

F.3.3 Plotting SINR with Varying Interference Power

```
clear
clc
n = 1;
sig = 0;
int = 0;
sinr = 0;
std_sig = 0;
std_int = 0;
std_sinr = 0;
SNR = 20;
for intpwr = 0:0.1:1
        temp_sig = 0;
        temp_int = 0;
        for i = 1:100
            [S, I, l] = corr_sir(SNR, intpwr);
            temp_sig(i) = S;
```

```
            temp_int(i) = I;
        end
        sig(n) = mean(temp_sig);
        int(n) = mean(temp_int);
        sinr(n) = sig(n) - int(n);

        std_sig(n) = std(temp_sig);
        std_int(n) = std(temp_int);
        std_sinr(n) = std_sig(n) - std_int(n);

        n = n+1;
end

intpwr = 0.0:0.1:1.0;
fig = figure(1);
plot(intpwr, sinr, 'o-b', 'LineWidth', 2.0, 'MarkerFace', 'b', 'MarkerSize', 10)
hold on
plot(intpwr, sinr + std_sinr, '^k', 'MarkerFace', 'k')
plot(intpwr, sinr - std_sinr, '^k', 'MarkerFace', 'k')

xlabel('Interference Power (W)', 'FontSize', 20, 'FontWeight', 'bold');
ylabel('Estimated SINR (dB)', 'FontSize', 20, 'FontWeight', 'bold');
set(fig, 'Position', [100 100 1000 640])
set(gca, 'FontSize', 18);
set(gca, 'Position', [0.075 0.1 0.875 0.875])
leg = legend('Estimated SINR', 'Standard Deviation of Estimate');

fig = figure(2);
plot(intpwr, sig, 'o-b', 'LineWidth', 2.0, 'MarkerFace', 'b', 'MarkerSize', 10)
hold on
plot(intpwr, sig + std_sig, '^k', 'MarkerFace', 'k')
plot(intpwr, sig - std_sig, '^k', 'MarkerFace', 'k')

xlabel('Interference Power (W)', 'FontSize', 20, 'FontWeight', 'bold');
ylabel('Estimated Signal Power (dBW)', 'FontSize', 20, 'FontWeight', 'bold');
set(fig, 'Position', [100 100 1000 640])
set(gca, 'FontSize', 18);
set(gca, 'Position', [0.085 0.1 0.875 0.875])
leg = legend('Estimated Signal Power', 'Standard Deviation of Estimate');
```

Acronyms

1xRTT 1 times Radio Transmission Technology

3GPP Third Generation Partnership Project

AI artificial intelligence

API application programming interface

ADC analog to digital converter

ASIC application-specific integrated circuit

AWGN additive white Gaussian noise

BER bit error rate

BT bandwidth-time product

DBPSK differential binary phase shift keying

CBDT case-based decision theory

CBR case-based reasoning

CDMA code division multiple access

CE cognitive engine

CES constant-elasticity-of-substitution

CR cognitive radio

CRC cyclic redundancy check

CTVR Centre for Telecommunications Value-Chain Research

CWT Center for Wireless Telecommunications

DAC digital to analog converter

DSA dynamic spectrum access

DSP digital signal processors

DSSS direct sequence spread spectrum

DTD document type definition

EDGE Enhanced Data Rates for GSM Evolution

EIRP effective isotropic radiated power

EV-DO Evolution-Data Optimized

FCC Federal Communications Commission

FDMA frequency division multiple access

FEC forward error correction

FER frame error rate

FFT fast Fourier transform

FHSS frequency hopping spread spectrum

FIFO first in first out

FIR finite impulse response

FM frequency modulation

FPGA field programmable gate arrays

FSF Free Software Foundation

GA genetic algorithm

GECCO Genetic and Evolutionary Computation Conference

GPRS General Packet Radio Service

GigE gigabit Ethernet

GPL general public license

GPP general purpose processors

GPU graphics processing unit

GNU GNU is not Unix

GSM global system for mobile communications

HMM hidden Markov models

HSPA High Speed Packet Access

IF intermediate frequency

IIR infinite impulse response

IRIS Implementing Radio in Software

ISI intersymbol interference

KUAR Kansas University Agile Radio

LTE Long Term Evolution

MAC Medium Access Control

MODM multiobjective decision making

MOGA multiobjective genetic algorithm

MRC maximal ratio combining

MSOPS multiple single objective Pareto sampling

NCO numerically controlled oscillator

OFDM orthogonal frequency division multiplexing

PSCR public safety cognitive radio

PDF probability density function

PHY physical

QA quality assurance

QoS quality of service

RF radio frequency

RC raised cosine

RRC root raised cosine

RSO repeated single objective

SCC Standards Coordinating Committee

SDR software defined radio

SIMD single instruction multiple data

SINR signal to interference plus noise ratio

SIR signal to interference ratio

SNR signal-to-noise ratio

SOAP Simple Object Access Protocol

SQL Structured Query Language

SR software radio

SSP subset-sum problem

TDMA time division multiple access

UMTS Universal Mobile Telecommunications System

USRP Universal Software Radio Peripheral

VHDL very high-speed integrated circuit hardware description language

VT Virginia Tech

WiMAX Worldwide Interoperability for Microwave Acess

WRAN wireless regional area networks

WSGA wireless system genetic algorithm

XML eXtensible Markup Language

About the Authors

Thomas W. Rondeau

Thomas W. Rondeau on the research staff of IDA's Center for Communications Research in Princeton, New Jersey. He received his B.S., graduating summa cum laude, and M.S. in electrical engineering from Virginia Tech in 2003 and 2006, respectively. He received his Ph.D. degree in electrical engineering from Virginia Tech in 2007. Upon completion of his degrees, he worked as a post-doctoral research engineer with the Centre for Telecommunications Value-Chain Research (CTVR) at Trinity College, Dublin, Ireland.

Rondeau has published over twenty papers on software defined and cognitive radios, including a chapter in the text book *Cognitive Radio Technologies* edited by Dr. Bruce Fette and a best paper award at the Software Defined Radio Forum's 2004 Technical Conference. Rondeau's doctoral dissertation was awarded both the Virginia Tech and Council of Graduate School's Distinguished Dissertation Award in mathematics, science, and engineering for 2007. The basis of this work has also been awarded one of the first U.S. patents on cognitive radio. He participates in the GNU Radio project where he has helped develop much of the digital communications capabilities including narrowband transmitters and receivers as well as an implementation of orthogonal frequency division multiplexing (OFDM). Rondeau has also had experience teaching digital communications to senior undergraduate students.

Tom's research interests include cognitive radio, software defined radio theory and implementation, artificial intelligence, signal processing, and software and programming practices. When not pursuing these interests, Tom can often be found reading and has a particular fascination with science and computing history and cultural trends.

Charles W. Bostian

Charles W. Bostian is Alumni Distinguished Professor of Electrical and Computer Engineering at Virginia Tech, where he has been a faculty member since 1969. Prior to joining the university, he served as a U.S. Army officer and worked briefly for Corning Glassworks. He holds B.S. (1963), M.S. (1964), and Ph.D. (1967) degrees from North Carolina State University. Since 1993, Bostian has served as the Director of the Virginia Tech Center for Wireless Telecommunications (CWT). He is also a member of *Wireless@Virginia Tech.*

In his career at Virginia Tech, Bostian has taught more than 4,000 students, and his teaching has been recognized by a number of awards, including ten certificates of teaching excellence and the William E. Wine Award for Excellence in Teaching. He is a four-time winner of the Eta Kappa Nu Outstanding Teaching Award and an elected member of the Virginia Tech Academy of Teaching Excellence. Bostian is the coauthor of two widely used textbooks, *Solid State Radio Engineering* and *Satellite Communications*, now in its second edition.

Bostian's primary research interests are in cognitive electronics and radio system design. Currently, he directs National Science Foundation (NSF), National Institute of Justice (NIJ) and Defense Advanced Research Projects Agency (DARPA) projects on cognitive radio. He has served on two international technology assessment panels sponsored by NSF and NASA, visiting many communications research centers in Europe and Japan. These panels produced two widely read reports that significantly influenced the direction of satellite communications research. One of these was republished in hard cover by Noyes as *Satellite Communications Systems and Technology*. He has authored or coauthored 45 journal and magazine articles and approximately 100 conference papers and presentations and contributed to the *Wiley Encyclopedia of Electrical and Electronics Engineering* and to *Cognitive Radio Technology* (Newnes, 2006).

Elected a Fellow of the IEEE in 1992 for contributions to and leadership in the understanding of satellite path radio wave propagation, Bostian is a former chair of the IEEE-USA Engineering R&D Policy Committee and served as Associate Editor for Propagation of *IEEE Transactions on Antennas and Propagation*. On leave during the 1989 calendar year, he was as an IEEE Congressional Fellow on the staff of U.S. Representative Don Ritter, working on legislative issues related to the American electronics industry and economic competitiveness. He served on the IEEE-USA Congressional Fellow Committee, helping to select and mentor other congressional fellows. He is a Fellow of the Radio Club of America.

In his off-duty hours, Bostian is a performing folk musician, playing hammered dulcimer and string bass with the band *Simple Gifts of the*

Blue Ridge. They have released four CDs, and the Canadian Broadcasting Corporation and Australian Public Radio have featured their music.

Index

Recent Titles in the Artech House Mobile Communications Series

John Walker, Series Editor

3G CDMA2000 Wireless System Engineering, Samuel C. Yang

3G Multimedia Network Services, Accounting, and User Profiles, Freddy Ghys, Marcel Mampaey, Michel Smouts, and Arto Vaaraniemi

802.11 WLANs and IP Networking: Security, QoS, and Mobility, Anand R. Prasad, Neeli R. Prasad

Achieving Interoperability in Critical IT and Communications Systems, Robert I. Desourdis, Peter J. Rosamilia, Christopher P. Jacobson, James E. Sinclair, and James R. McClure

Advances in 3G Enhanced Technologies for Wireless Communications, Jiangzhou Wang and Tung-Sang Ng, editors

Advances in Mobile Information Systems, John Walker, editor

Advances in Mobile Radio Access Networks, Y. Jay Guo

Applied Satellite Navigation Using GPS, GALILEO, and Augmentation Systems, Ramjee Prasad and Marina Ruggieri

Artificial Intelligence in Wireless Communications, Thomas W. Rondeau and Charles W. Bostian

Broadband Wireless Access and Local Network: Mobile WiMax and WiFi, Byeong Gi Lee and Sunghyun Choi

CDMA for Wireless Personal Communications, Ramjee Prasad

CDMA Mobile Radio Design, John B. Groe and Lawrence E. Larson

CDMA RF System Engineering, Samuel C. Yang

CDMA Systems Capacity Engineering, Kiseon Kim and Insoo Koo

CDMA Systems Engineering Handbook, Jhong S. Lee and Leonard E. Miller

Cell Planning for Wireless Communications, Manuel F. Cátedra and Jesús Pérez-Arriaga

Cellular Communications: Worldwide Market Development, Garry A. Garrard

Cellular Mobile Systems Engineering, Saleh Faruque

The Complete Wireless Communications Professional: A Guide for Engineers and Managers, William Webb

EDGE for Mobile Internet, Emmanuel Seurre, Patrick Savelli, and Pierre-Jean Pietri

Emerging Public Safety Wireless Communication Systems, Robert I. Desourdis, Jr., et al.

The Future of Wireless Communications, William Webb

Geographic Information Systems Demystified, Stephen R. Galati

GPRS for Mobile Internet, Emmanuel Seurre, Patrick Savelli, and Pierre-Jean Pietri

GPRS: Gateway to Third Generation Mobile Networks, Gunnar Heine and Holger Sagkob

GSM and Personal Communications Handbook, Siegmund M. Redl, Matthias K. Weber, and Malcolm W. Oliphant

GSM Networks: Protocols, Terminology, and Implementation, Gunnar Heine

GSM System Engineering, Asha Mehrotra

Handbook of Land-Mobile Radio System Coverage, Garry C. Hess

Handbook of Mobile Radio Networks, Sami Tabbane

High-Speed Wireless ATM and LANs, Benny Bing

Interference Analysis and Reduction for Wireless Systems, Peter Stavroulakis

Introduction to 3G Mobile Communications, Second Edition, Juha Korhonen

Introduction to Communication Systems Simulation, Maurice Schiff

Introduction to Digital Professional Mobile Radio, Hans-Peter A. Ketterling

Introduction to GPS: The Global Positioning System, Ahmed El-Rabbany

An Introduction to GSM, Siegmund M. Redl, Matthias K. Weber, and Malcolm W. Oliphant

Introduction to Mobile Communications Engineering, José M. Hernando and F. Pérez-Fontán

Introduction to Radio Propagation for Fixed and Mobile Communications, John Doble

Introduction to Wireless Local Loop, Second Edition: Broadband and Narrowband Systems, William Webb

IS-136 TDMA Technology, Economics, and Services, Lawrence Harte, Adrian Smith, and Charles A. Jacobs

Location Management and Routing in Mobile Wireless Networks, Amitava Mukherjee, Somprakash Bandyopadhyay, and Debashis Saha

Mobile Data Communications Systems, Peter Wong and David Britland

Mobile IP Technology for M-Business, Mark Norris

Mobile Satellite Communications, Shingo Ohmori, Hiromitsu Wakana, and Seiichiro Kawase

Mobile Telecommunications Standards: GSM, UMTS, TETRA, and ERMES, Rudi Bekkers

Mobile Telecommunications: Standards, Regulation, and Applications, Rudi Bekkers and Jan Smits

Multiantenna Digital Radio Transmission, Massimiliano Martone

Multiantenna Wireless Communications Systems, Sergio Barbarossa

Multipath Phenomena in Cellular Networks, Nathan Blaunstein and Jørgen Bach Andersen

Multiuser Detection in CDMA Mobile Terminals, Piero Castoldi

OFDMA for Broadband Wireless Access,

Personal Wireless Communication with DECT and PWT, John Phillips and Gerard Mac Namee

Practical Wireless Data Modem Design, Jonathon Y. C. Cheah

Prime Codes with Applications to CDMA Optical and Wireless Networks, Guu-Chang Yang and Wing C. Kwong

QoS in Integrated 3G Networks, Robert Lloyd-Evans

Radio Engineering for Wireless Communication and Sensor Applications, Antti V. Räisänen and Arto Lehto

Radio Propagation in Cellular Networks, Nathan Blaunstein

Radio Resource Management for Wireless Networks, Jens Zander and Seong-Lyun Kim

Radiowave Propagation and Antennas for Personal Communications, Third Edition, Kazimierz Siwiak and Yasaman Bahreini

RDS: The Radio Data System, Dietmar Kopitz and Bev Marks

Resource Allocation in Hierarchical Cellular Systems, Lauro Ortigoza-Guerrero and A. Hamid Aghvami

RF and Baseband Techniques for Software-Defined Radio Peter B. Kenington

RF and Microwave Circuit Design for Wireless Communications, Lawrence E. Larson, editor

Sample Rate Conversion in Software Configurable Radios, Tim Hentschel

Signal Processing Applications in CDMA Communications, Hui Liu

Smart Antenna Engineering, Ahmed El Zooghby

Software Defined Radio for 3G, Paul Burns

Spread Spectrum CDMA Systems for Wireless Communications, Savo G. Glisic and Branka Vucetic

Technologies and Systems for Access and Transport Networks, Jan A. Audestad

Third Generation Wireless Systems, Volume 1: Post-Shannon Signal Architectures, George M. Calhoun

Traffic Analysis and Design of Wireless IP Networks, Toni Janevski

Transmission Systems Design Handbook for Wireless Networks, Harvey Lehpamer

UMTS and Mobile Computing, Alexander Joseph Huber and Josef Franz Huber

Understanding Cellular Radio, William Webb

Understanding Digital PCS: The TDMA Standard, Cameron Kelly Coursey

Understanding GPS: Principles and Applications, Second Edtion, Elliott D. Kaplan and Christopher J. Hegarty, editors

Understanding WAP: Wireless Applications, Devices, and Services, Marcel van der Heijden and Marcus Taylor, editors

Universal Wireless Personal Communications, Ramjee Prasad

WCDMA: Towards IP Mobility and Mobile Internet, Tero Ojanperä and Ramjee Prasad, editors

Wireless Communications in Developing Countries: Cellular and Satellite Systems, Rachael E. Schwartz

Wireless Communications Evolution to 3G and Beyond, Saad Z. Asif

Wireless Intelligent Networking, Gerry Christensen, Paul G. Florack, and Robert Duncan

Wireless LAN Standards and Applications, Asunción Santamaría and Francisco J. López-Hernández, editors

Wireless Technician's Handbook, Second Edition, Andrew Miceli

For further information on these and other Artech House titles,including previously considered out-of-print books now available through our In-Print-Forever® (IPF®) program, contact:

Artech House
685 Canton Street
Norwood, MA 02062
Phone: 781-769-9750
Fax: 781-769-6334
e-mail: artech@artechhouse.com

Artech House
16 Sussex Street
London SW1V 4RW UK
Phone: +44 (0)20 7596-8750
Fax: +44 (0)20 7630-0166
e-mail: artech-uk@artechhouse.com

Find us on the World Wide Web at: www.artechhouse.com